Agro-Economic Risks of *Phytophthora* and an Effective Biocontrol Approach

Edited by

Waleed Mohamed Hussain Abdulkhair

Published in London, United Kingdom

IntechOpen

Supporting open minds since 2005

Agro-Economic Risks of *Phytophthora* and an Effective Biocontrol Approach
http://dx.doi.org/10.5772/intechopen.95679
Edited by Waleed Mohamed Hussain Abdulkhair

Contributors
Joseph Adomako, Stephen Larbi-Koranteng, Frederick Kankam, Stephen Yeboah, Yaw Danso, Jerry Fenteng Asamoah, Kennedy Agyeman, Eric Owusu Danquah, Patricia Oteng-Darko, Erasmus Tetteh Nartey, Brahim Oubaha, Abdellah Ezzanad, Hernando Jose Bolívar-Anillo, Benedicta Nsiah Frimpong, Samuel Oteng Ampadu, Allen Oppong, Isaac Nunoo, Lydia Brobbey, Pooja Vajpayee, Kuldeep Kr. Yogi, Carlos Garrido, Victoria E. González-Rodríguez, Giovanna Reyes Almeida, Javier Moraga, Inmaculada Izquierdo-Bueno, Jesús Manuel Cantoral Fernández, Maria Carbú, Waleed Mohamed Hussain Abdulkhair

First published in London, United Kingdom, 2021 by IntechOpen
IntechOpen is the global imprint of INTECHOPEN LIMITED, registered in England and Wales, registration number: 11086078, 5 Princes Gate Court, London, SW7 2QJ, United Kingdom
Printed in Croatia

British Library Cataloguing-in-Publication Data
A catalogue record for this book is available from the British Library

Additional hard and PDF copies can be obtained from orders@intechopen.com

Agro-Economic Risks of *Phytophthora* and an Effective Biocontrol Approach
Edited by Waleed Mohamed Hussain Abdulkhair
p. cm.
Print ISBN 978-1-83969-653-4
Online ISBN 978-1-83969-654-1
eBook (PDF) ISBN 978-1-83969-655-8

For EU product safety concerns:
IN TECH d.o.o., Prolaz Marije Krucifikse Kozulić 3, 51000 Rijeka, Croatia, info@intechopen.com or visit our website at intechopen.com.

Meet the editor

Waleed Abdulkhair is an Associate Professor of Microbiology, Department of Microbiology, Egyptian Drug Authority. He has participated in several international, national, and local conferences on biological and chemical control of plant pathogens. He has published many scientific papers on biological control of plant pathogenic bacteria and fungi in specialized international scientific journals. In 2015, he and an integrated team obtained a patent from the Saudi Patent Office to isolate and identify a novel bacterial strain of the genus *Streptomyces* characterized by its superior ability to destroy plant pathogenic fungi, especially *Fusarium oxysporum lycopersici*, which infects and destroys tomato plants. He has co-authored several books on plant pathogens. He is also the editor of a book on yeast and its applied medical roles.

Contents

Preface

Phytophthora is a microbial genus belonging to a fungi-like group of oomycetal phytopathogens. This genus has many species that affect different parts of the plant, especially the root and stem base. The diseases of *Phytophthora* are usually accompanied by definite symptoms such as wilting and foliage discoloration. *Phytophthora* is widely prevalent worldwide because it can survive under extreme conditions and different habitats like soil and plant debris.

Microbial infection with *Phytophthora* spp., one of the more destructive phytopathogens, is a common reason for loss of economic crops. The agronomic countries depend mainly on exportation of essential plants like fruits, vegetables, and cereals. Thus, damage to these plants leads to economic loss and food security threats. Various species of *Phytophthora*, including *P. cinnamomi*, *P. cryptogea*, *P. citricola*, *P. cactorum*, and *P. cambivora*, damage different plants including ornamental horticulture plants. *Phytophthora* infects the aerial parts of the plant, targeting the root and stem base. However, some species of *Phytophthora*, such as *P. ilicis*, *P. ramorum*, and *P. kernoviae*, affect only the aerial parts of plants.such as *P. ilicis*, *P. ramorum*, and *P. kernoviae*.

Discoloration and decay are symptoms of infection of roots and stem bases. Foliar symptoms do not appear until most of the plant's roots have been killed, or microbial invasion of the stem base has been carried out. The infected foliar parts of plants like conifers are usually associated with clear symptoms such as wilting, shedding of leaves, dieback, and a gradual fading of color. *Phytophthora* infection leads to retarding of water and nutrient delivery to the stem, therefore resulting in the symptoms mentioned previously. Extreme conditions like drought will hasten their onset. As such, constant examination of roots and stem bases is essential because the foliar symptoms never appear at the early stages of *Phytophthora* infection of the roots. Some other foliar symptoms are represented as wet, dark brown to black spots on the leaves and stems causing wilting and aerial dieback but no stem or root death.

Although *Phytophthora* belongs to a fungi-like group called oomycetes, it is more closely related to algae. *Phytophthora* is a motile microorganism that migrates via tiny spores called zoospores, which quickly move in water and attract to the roots due to chemical exudates. Poor discharging or retaining of water inside the plant leads to worse disease because zoospores require water to survive and move. Other long-living spores produced by the *Phytophthora* species are called oospores and chlamydospores, which are liberated after roots decay in the soil causing contamination even for compost and standing areas. But oospores are more dangerous than chlamydospores because they can survive outside the host. Unfortunately, the fungicides used for *Phytophthora* control are fungistatic not fungicidal, so disease progression takes place after fungicide decays. Other than fungicides, there are various chemical disinfectants and non-chemical methods like steaming, solarization, and bio-fumigation used for *Phytophthora* control.

The final and radical solution to the problem of *Phytophthora* infection of plants is based on two axes. The first axis is prophylactic, that is, prevention of *Phytophthora* infestation via continuous disinfection of irrigation water sources, preventing sewage from reaching the soil, purifying the soil continuously from weeds and harmful plants, and spraying pesticides and fungicides in reduced doses. The second axis depends on the treatment in the event of any infection, whether to the roots, stems, or foliar parts of the plant, by spraying fungicides in intensive doses with re-examination of the tissues of all plant parts to ensure plant safety and to stop disease spread.

Advisements given to farmers in the form of preventive measures include draining any open soil and standing water, ensuring water irrigation is free from *Phytophthora*, covering water reservoirs to prevent contamination, filtrating and sterilizing water, using disposable tools like pots and trays or at least carefully disinfecting reused ones, and taking only healthy cuttings at propagation. Those individuals importing plants should inspect all plants and remove any that show signs of infection.

This book is divided into three sections. Section 1, "*Phytophthora* Threats on Economic Plants in Egypt and Ghana," includes Chapter 1 "*Phytophthora* spp.: Economic Plant Threats in Egypt" and Chapter 2 "*Phytophthora* Diseases in Ghana: Threats, Management Options and Prospects." Section 2, "Biocontrol of *Phytophthora* Infections," includes Chapter 3 "Plant Beneficial Microbes Controlling Late Blight Pathogen, *Phytophthora infestans*" and Chapter 4 "Endophytic Microorganisms as an Alternative for the Biocontrol of *Phytophthora* spp.". Section 3, "Prevalence and Recognition of *Phytophthora*," includes Chapter 5 "*Phytophthora* Diseases Prevalence, Its Effects and Controls in Ghana" and Chapter 6 "Recognition and Early Stage Detection of *Phytophthora* in a Crop Farm Using IoT".

Waleed Mohamed Hussain Abdulkhair
Microbiology Department,
Egyptian Drug Authority,
(National Organization for Drug Control and Research, formerly),
Giza, Egypt

Section 1

Phytophthora Threats on Economic Plants in Egypt and Ghana

Chapter 1

Phytophthora spp.: Economic Plant Threats in Egypt

Waleed Mohamed Hussain Abdulkhair

Abstract

The potato crop is exposed to infection with many fungal diseases including late blight, caused by *Phytophthora infestans*. The control of late blight disease requires an integrated management approach represented in cultivation control, plant resistance, and fungicide control. The citrus plants are infected by *Phytophthora nicotianae* that is causing root rot disease in Egypt. Three species of *Phytophthora* responsible for infection of citrus plants; *P. nicotianae*, *P. citrophthora*, and *P. palmivora*. Other pathogens associate *P. nicotianae* and form complexes or coinfection that release different diseases for citrus plants such as gummosis, *Phytophthora–Diaprepes* complex (PDC), and Huanglongbing syndrome (HLBS).

Keywords: oomycetes, pathogenicity, coinfection, citrus plants, potatoes

1. Introduction

Plants are exposed to various biotic and abiotic stresses that affect their growth and yield. Biotic stress is represented by different microbial diseases that infect the plant. Therefore, the control of biotic stress is very urgent to maintain healthy crops Egypt suffers from accelerated population growth, which hinders economic development and constitutes an explicit threat to food security if it is not decisively controlled. FAO reported that "In the next decades, Africa will suffer from severe decrease of crops due to water deficiency, adverse weather events, pests and other factors, which lead to famines and drought".

Crops are exposed to microbial infections particularly fungal infections including *Phytophthora* infection as one of the main destructive phytopathogens. Among these crops, citrus is exposed to damage and subsequently high economic loss due to *Phytophthora* infection. – spp. infect different parts of citrus plants causing various diseases such as damping-off of seedlings, fibrous root rot, crown rot, and gummosis [1]. Ten species of genus *Phytophthora* are known to infect citrus plants around the world causing serious diseases such as gummosis, and root and fruit rots [2]. Chemical fungicides are one of the control measures, which provide an effective result, but their effects are not eco-friendly due to their harmful effects. Different fungicides, including Fosetyl-Al and Metalaxyl-M, may be used separately or by alternation to reduce the development of fungicide resistance [3]. Metalaxyl-M is usually used in East African countries to control the *Phytophthora* infection prevalence in potatoes [4, 5]. Climatic conditions play an important role in the prevalence of *Phytophthora* infection in citrus plants; i.e., the root infections caused by *P. citrophthora* are severe during spring and very mild during winter, on the other hand, the root infections caused by *P. nicotianae* are severe during summer and early autumn and very mild during winter [6].

Potato is one of the economic crops which are important for Egyptian exportation. This strategic and economic crop is exposed to complete damage by late blight disease caused by *P. infestans*, which is considered the most common disease for both potatoes and tomatoes worldwide [7–10]. Control of *Phytophthora* infection depends mainly on the contentious elimination of spoiled piles of potatoes and using a convenient means for both harvesting and storage [11]. Although different fungicides are used in the control of late blight, the phenyl amide fungicides such as Ridomil still the most effective ones [12, 13]. Nevertheless, the use of Ridomil leads to resistance development with frequent usage, so the strategy of Ridomil practice should be modified; i.e., the potatoes should be treated with Ridomil (0.75 kg ha^{-1}) followed by another dose (1.5 kg ha^{-1}) to prevent both formations of late blight disease and resistance development [14–16]. In such a way, using fungicides as an effective treatment for late blight disease increases the productivity of potatoes by as much as 60% and therefore decreases the economic loss [17]. In this chapter, we will elucidate the harmful effects of *Phytophthora* spp. on crops in Egypt, and the followed recent methods to control its diseases to maintain sustainable agriculture and a strong economy, therefore.

2. *Phytophthora*

The phytopathogenic fungus *Phytophthora* spp. is unrelated to true fungi and belongs to oomycetes. This oomycete has mycelia which released branched sporangiophores, which produce lemon-shaped sporangia at their tips. This oomycete is characteristic of swellings produced by sporangiophores at the places of sporangia formation [18, 19]. Different species are belonging to the genus *Phytophthora* including *Phytophthora infestans* which have two mating types (A1 and A2) which are required to produce sexual spores known as oospores. *P. infestans* is the causative pathogen of potato late blight disease [18–20]. Oospores are the main reason for new strains' release of *P. infestans* worldwide, which are more deleterious than the old strains. Sexual reproduction begins with the growth of the two mating types adjacently, where the globose oogonium is developed above the antheridium due to the growth of female hypha through the young antheridium, which in turn fertilizes the oogonium to develop a thick-walled and hardy oospore. The germination of oospores is carried out by a germ tube that produces sporangia, which germinate entirely by releasing three to eight zoospores at temperatures up to 12 or 15°C, whereas above 15°C sporangia may germinate directly by producing a germ tube [19].

3. *Phytophthora* diseases in Egypt

Egypt is an agronomic country that depends on the exportation of crops to develop its economy. Therefore, all agronomic pests including *Phytophthora* spp. which destroy the crops and other plants are threats and lead to decline of the economy. Potatoes and citrus plants are occupied the first ranks of Egyptian agronomic crops which export to different countries over the world, especially the European Union and Arabic Gulf countries. In Egypt, potatoes and citrus plants are usually attacked by *P. infestans* and *P. nicotianae*, respectively. These phytopathogenic fungi are very deleterious due to their severe effects and resistance ability for many fungicides. So, the control of these pathogens is a very urgent approach to maintain healthy crops, food security, and a strong economy. In the last 5 years, the Egyptian government has succeeded to render the orange crop is the first exported citrus plant worldwide due to new strategies and sustainable agriculture programmes which are applied intensively.

4. *P. infestans*

4.1 Life cycle

Temperature plays an important role in its life cycle, where sporangia of *P. infestans* can germinate by a single germ tube at 13–21°C, and the emergence of late blight is accompanied by a moderate temperature (10–16°C) and high relative humidity [21, 22]. The fungal growth is inhibited at 30°C or more, and it can further sporulate when the temperature be relatively moderate with high relative humidity near 100% [19]. The emergence of lesions is the main symptom of late blight disease. These lesions have appeared as small circular or irregular with light to dark green color, and they are also water-soaked [22]. These lesions are appeared on both lower and upper leaves, where unfavorable and favorable climate conditions are present, respectively [22]. At high humidity, the lesions are fast enlarged with the formation of brown areas with the absence of clear borders. At these borders, a white zone (3–5 mm) of downy mildew growth has appeared on the lower sides of the leaves, and then all leaves are infected and die [19]. Therefore, there is a relationship between late blight infection and humidity, where the infection and symptoms appearance are accompanied by high relative humidity for a minimum of 7–10 hours. The late blight infection is distributed via spores scattering by wind and rain until reach healthy plants where the disease cycle begins again. The life cycle of *P. infestans* depends on two main environmental factors; humidity and temperature. The moderate temperature and high relative humidity intensively allow to formation and distribution of late blight infection. Many life cycles maybe happen in one season, and subsequently, the oomycetal infection is more prevalent even in the soil as a carrier for spores. The late blight disease can completely damage the potatoes within a few days or a few weeks when favorable environmental conditions are provided. On the other hand, late blight disease is attenuated and disappeared with dry weather, where the oomycetal activity is stopped and the existing lesions do not enlarge and turn black, curl, and wither [19].

The late blight infection begins with close contact with sporangia and tubers. Usually, late blight infection happens when sporangia are penetrating the tubers of potatoes. Developing and mature tubers are a target for sporangia; however, mature ones are a more favorable targets because they can cause cracks in the soil and give sporangia ready access. The moist soil accelerates the infection of tubers, which are usually appeared as irregular dark purple or brownish blotches. When the infected tuber is torn off, water-soaked, dark, reddish-brown infected tissues have appeared, and the oomycetal infection may be deeply extended into 5–15 mm in the tuber flesh. The coinfection may happen for the infected tubers, where other fungal and bacterial infections are present. The coinfection causes soft rots which characterize by a putrid taste and offensive odor [19, 22].

There are two types of sporangia germination of *P. infestans*; direct germination by a germ tube or indirect germination by zoospores formation. Secondary sporangia may be produced by germ tubes, and germinate at 7–13°C with the presence of free water on leaves, and form 8–12 motile zoospores per sporangium, which swim in the water and attach to the leaf surface and infect the plant. Zoospores liberate the germ tubes and by which infect the leaves either through stomata or by direct penetration, where the mycelia grow and release long-curled haustoria between the cells. The second phase of the life cycle starts in the moist soil bearing sporangia. Zoospores profusely germinate and penetrate the tubers through the wounds [21, 22].

4.2 Control

The control of late blight infection depends on different management approaches including cultivation control, plant resistance and fungicides control.

4.2.1 Cultivation control

Control of late blight disease can be achieved by using good agricultural practices as the most common control technique including using healthy seeds, elimination of all infected potatoes, avoidance of frequent or night-time overhead irrigation, using more conventional methods for both harvesting and storage, and using more effective fungicides and herbicides to destroy the dropped infected parts of potato plant [23]. As mentioned above, the prevalence of late blight disease depends on providing adequate weather conditions (moderate temperature and high relative humidity) despite these conditions being beyond the control. Nevertheless, some measures should be applied such as the selection of proper fields and taking into account the appropriate method, time and amount of irrigation. The proper field should have good water seepage, high drainage, free from inoculum sources (infected potatoes), and also free from the weeds which restrict the air movement within the canopy and block the access of fungicides to the leaves and tubers of potatoes, and alternative late blight hosts like hairy nightshade that supports the disease prevalence [21, 22]. Sprinkler irrigation is a more adequate method for potato cultivation because it does not provide sufficient humidity that allows fungal growth and late blight disease incidence. It is preferable to grow potatoes in rows parallel with the winds to improve the air circulation and therefore foliage dryness. Conscious cultivation of potatoes requires rooting of the vines before 2 weeks of harvest, and then spraying the foliage with fungicides to kill living late blight spores. The healthy tubers should be dried during and after storage, while the infected tubers should be separated from the healthy ones [24].

4.2.2 Plant resistance

The plant resistance to late blight disease is another means for control that saves us from using fungicides and subsequently prevents or diminishes fungicide resistance. The plant resistance for late blight disease is an eco-friendly means, so the modern approach of potato cultivation depends on using resistant cultivars, which vary among each other in the resistance rate to late blight disease. Therefore, the pursuit of late blight resistance is a very urgent matter, so a lot of potato cultivars are genetically engineered to be resistant. Nevertheless, these genetically engineered cultivars can be destroyed by other new strains of *P. infestans* because the resistance is encoded by a single gene. So, using of polygenic (durable) resistant cultivars is very helpful to allow the production of healthy potatoes free from any deposits of fungicides. Although plant resistance technique is very effective for control of late blight disease, avoidance of cultivation inadequate weather for late blight emergence is required; as well as, frequent spray with an effective fungicide to completely prevent the growth of *P. infestans* [25].

4.2.3 Fungicides control

Despite the two previous effective methods in controlling the spread of late blight that infects potato plants, using fungicides are very essential to completely prevent or eradicate the late blight disease. The fungicides are also used as prophylactic agents, where they are sprayed on healthy potatoes to prevent any growth chance of *P. infestans*. The use of fungicides must be frequent and periodically because they may be broken down by weather factors. There are two types of fungicides according to their mobility named protectant and penetrant. The protectant fungicides are usually used before the incidence of late blight infection because the already infected potatoes never get rid of the symptoms of late blight and the damaged tissues are never repaired. The penetrant fungicides are effectively used after the incidence of late blight infection because they can kill the fungus and stop the prevalence of late blight infection. Both protectant and penetrant fungicides have broad-spectrum and systemic action, so they are powerful chemical control agents [26].

5. *P. nicotianae*

5.1 Description

The main pathogen of citrus plants in Egypt is *P. nicotianae* or also called *P. parasitica*. This phytopathogenic oomycete was firstly isolated from tobacco in Indonesia in 1896s. The oospores are produced from both antheridia and septate oogonia. The spherical, ovoidal, or ellipsoidal sporangia are present with one to three sharp papillae. The intercalary or terminal spherical chlamydospores and arachnoid mycelia are also present. Although *P. nicotianae* can infect around 90 different plant families, the citrus plants are still the main target of it. So, this pathogen causes a high loss of citrus plants to reach to 15% in Egypt and subsequently negatively affects the Egyptian exports from citrus plants and in turn leads to significant economic loss. The isolation of *P. nicotianae* from the soil can be carried out by different methods, but the baiting method is the most common and effective one. *P. nicotianae* can be identified either by classical methods depending on morphology determination or by genetical methods depending on SSCP fingerprinting and ITS sequencing. The eradication of *P. nicotianae* depends on the same methods mentioned above especially using effective fungicides such as Metalaxyl-M and phosphonate.

5.2 Pathogenicity spectrum

The genus *Phytophthora* has a wide array of species that reaches 120 due to the improvement of identification tools and methods, a wide survey of natural habitats, and reports of new diseases [27, 28]. This blossoming may increase in the next years where it has been reached 600 species [29]. Therefore, the host plants of *Phytophthora* genus are also extending to be about 4400 hosts [30]. The pathogenicity spectrum is differed from one species to another and subsequently influence on agronomic productivity. *P. nicotianae* is characterized by its wide pathogenicity spectrum against 90 families of plants particularly citrus ones and causes a high loss of productivity. *P. nicotianae* is widely distributed worldwide especially in temperate countries like Egypt and infects a considerable number of plants causing huge economic loss [31, 32]. This pathogen causes many plant diseases including brown rot, foot rot, root rot, gummosis, and black shank of tobacco [33].

5.3 Pathogenicity behavior

The phytopathogenic *P. nicotianae* usually infects the roots of different plants and other parts such as leaves, stems, and fruits and causes crown rot disease. The pathogenicity behavior of *P. nicotianae* is hemibiotrophic; i.e., the pathogenicity is accomplished through two steps; the first one is called biotrophy that implies the pathogen intimately contacts with the healthy tissues at the early stages of infection, and the second step is called necrotrophy in which the pathogen penetrates the tissues and profusely grows and causing tissue wilting and death. *P. nicotianae* has sporangiophores that bear multinucleate sporangia, which directly germinate in the proper weather and produce wall-less zoospores which are uninucleate and possess two flagella to can migrate until contact with the host tissues. Once contact is done, zoospores form a cell wall and cysts, which in turn germinate to form germ tubes by which they can penetrate the plant tissues [34, 35]. Moreover, *P. nicotianae* can reproduce sexually by formation thick-walled oospores as a net result of male and female gametangia fusion. Sexual reproduction leads to high genetic variation which in turn leads to releasing of novel pathogenicity and virulence factors. The main habitat of oospores is the soil where they can persist for several years until germination and formation of germ tube that penetrates the host tissues and causes the diseases.

Under hard extreme environmental conditions, *P. nicotianae* produces thick-walled asexual structures called chlamydospores, which persist in the soil very long time reach to several years. Chlamydospores can actively germinate at moderate and high temperatures, while they are dormant in low temperatures as the same with oospores. Therefore, *P. nicotianae* can tolerate the hard weather of the winter by both oospores and chlamydospores which be dormant in rhizospheres of host plants. Some animals like termites and snails are good vectors for both oospores and chlamydospores, which survive in their gastrointestinal tracts and feces [36]. It is argued that, *P. nicotianae* can reproduce sexually and asexually according to weather conditions; i.e., under proper conditions, they usually reproduce by sexual propagules, but under unfavorable conditions, they tend to asexual reproduction to tolerate the hard conditions. So, the life cycle of *P. nicotianae* includes both sexual and asexual propagules [37].

5.4 Coinfection

The coinfection implies the association of *P. nicotianae* as a soil-borne pathogen with other pathogens which together infect the host-plant tissues and cause unprecedented diseases such as the new citrus diseases, which are somewhat incurable diseases.

5.4.1 Phytophthora *gummosis*

Citrus plants represent a high economic value in Egypt, so they must be protected from all destructive pathogens especially *P. nicotianae*. Citrus plants may usually be infected by 10 species of *Phytophthora*, the severe diseases are incident by only three species namely *P. nicotianae*, *P. citrophthora* and *Promecotheca palmivora*, but *P. nicotianae* is the most destructive one not only in Egypt but all over the world where it causes root rot, foot rot, and gummosis [38]. The coinfection is usually accompanied by the life cycle of *P. nicotianae*, where it is associated with *P. palmivora* in warm countries like Florida or Southern Asia, and it is associated with *P. citrophthora* in the Mediterranean countries and causing branched cankers [39, 40]. *P. citrophthora* is active in spring, while *P. nicotianae* is active in the summer and early autumn causing root infection. Therefore, *P. nicotianae* requires a high temperature more than *P. citrophthora* for growth and activity. *P. nicotianae* is a dominant phytopathogenic fungus in Brazil, Egypt, South Africa, and Tunisia. Fungicides, resistant plants, and sanitary practices are good measures for the control of *P. nicotianae* [41].

5.4.2 Phytophthora*:* Diaprepes *complex (PDC)*

Diaprepes is the polyphagous root weevil found in citrus areas due to the distribution of infected nursery stock. The roots of citrus plants are being infected by the larvae of the weevil which are feeding on root tissues until be dilapidated and then be died. On the other hand, the infection with ***Diaprepes*** is followed by the infection with ***P. nicotianae*** through the injured roots predisposed to ***Phytophthora*** infections, and PDC is formed, which is an incurable disease that is very difficult to control [42].

5.4.3 Huanglongbing syndrome (HLBS)

There is another coinfection include *P. nicotianae* called HLBS that is more incurable than PDC. The name of this disease returns to the citrus greening or yellow dragon disease which is the oldest citrus disease worldwide. Although, this disease is widely distributed in most citrus areas over the world, the Mediterranean Basin region, Australia and Japan are less affected [43]. HLBS is usually transmitted by both psyllid vectors and grafting. Three varieties are belonging to unculturable Gram-negative bacterium; *Candidatus liberibacter*. The first variety was isolated from Africa; *C. liberibacter africanus* and designated as (CaLaf), the second variety was isolated from Asia; *C. liberibacter asiaticus* and designated as (CaLas), and the third and last variety was isolated from America; *C. liberibacter americanus* and designated as (CaLam) [44]. The effect of this disease is very severe because it causes vascular decline, reduces both fruit size and quality, and completely kills the trees [45]. The seriousness of this disease returns to no resistant plants are found and also no efficient management program is currently available, so the plant losses may reach 100% locally [46]. Moreover, the pathogenic bacteria attack all parts of the host plant and cause complete damage particularly if the coinfection included *P. nicotianae* is incident. Accordingly, the control of this disease mainly depends on the eradication of psyllid vectors and following the sanitary practices including removal of infected trees, which may be avoided by following a good nutrition program. Furthermore, effective mefenoxam-based fungicides can be used as helpful agents to get a ride of this dreaded plant disease. This control strategy has two disadvantages: high cost and inefficiency with mild or long persistence [3, 47].

6. Methodology

6.1 Isolation of pure cultures of *Phytophthora* spp.

Isolation of *Phytophthora* spp. is usually carried out by using Rye B Agar medium [48] (60 g rye grain, 20 g glucose/sucrose, 15 g agar, and 1.0 L distilled water, pH 6.8 ± 0.2). The rye grains were soaked in distilled water and left at room temperature for 36 hours. The supernatant was poured off and kept in a separate vessel, while the distilled water was added to the settled grains and heated in a water bath at 50°C for 3.0 hours. The mixture was filtered through a sieve and the grains were discarded. The filtrate was mixed with the kept supernatant, glucose/sucrose, and agar, and the distilled water was added until the volume reached 1.0 L, and then agitated thoroughly. The mixture was autoclaved at 121°C for 20 minutes. The autoclaved medium was left to be cooled at 45°C, and 3 ml of a stock rifamycin (30 μg/ml) and 1 ml of stock piramycin (10 μg/ml) were added and mixed thoroughly.

The infected part of plant with *Phytophthora* spp. was cut into small fragments, which were placed into the healthy part susceptible to infection. The two parts were fastened back together with a rubber band and incubated at 16°C for 4–9 days. The infected part was cut into small fragments by a sterile scalpel. These fragments were disinfected in 70% ethanol for 15–20 seconds and then de-aerate in 0.1% sublimate of $HgCl_2$ for 30–45 seconds, or in 1% sodium hypochlorite for 180 seconds. The disinfected fragments were rinsed three times with sterile distilled water, and placed onto the agar surface of Rye B medium containing antibiotics and maintained in dark at 16°C for 10 days. The colonies of *Phytophthora* were examined microscopically by a lens (400×). The sporangia of *Phytophthora* appeared in different shapes according to the species [49].

6.2 Storage of *Phytophthora* spp.

6.2.1 Using agar slopes under paraffin oil or water

The pure culture of *Phytophthora* spp. was grown on Rye B Agar medium for 2 weeks. The agar surface loaded by the growth of *Phytophthora* spp. was flooded with sterile paraffin oil or water and stored at 4–7°C. The sub-culture must carry out at least once every 3 years.

6.2.2 Using liquid nitrogen

The pure culture of *Phytophthora* spp. was grown on Rye B Agar medium for 10–14 days at 16°C. The growth was cut into discs by a sterile cork-borer, and placed in 1.5 ml cryovials containing a sterile 15% dimethyl sulfoxide solution (DMSO), and mixed thoroughly until be immersed. The temperature of cryovials containing *Phytophthora* discs was decreased gradually (−1°C/min) for at least 4 hours at −70°C using a specified device for this purpose. The frozen growth was transferred into liquid nitrogen (−196°C). Due to the toxic effect of DMSO on the growth of *Phytophthora* spp., the growth discs must be quickly thawed and rinsed carefully and thoroughly with sterile distilled water when out of the liquid nitrogen.

6.3 Preparation of *Phytophthora* inoculum

The mycelial growth of *Phytophthora* spp. was placed on the agar surface of Rye B Agar medium between two slices (1 cm) of susceptible host plant for infection. The two slices never completely cut off from each other to be closely attached with the moist medium to enhance the pathogen proliferation. The agar plates were incubated at 16°C for a week at high relative air humidity (80–100%) until thick mycelia appeared on the upper surface of the top slice. The sporangia of *Phytophthora* were collected from the mycelia by a brush and washed with deionized water. The hemocytometer was used to prepare the required inoculum concentration (50 sporangia μl^{-1}). The inoculum was left at 7°C for 2 hours, and at room temperature for half an hour to increase the liberation of zoospores from sporangia. During the test, the inoculum should be constantly and gently agitated to prevent sporangia sedimentation and therefore zoospores accumulation that provides an adverse event on the solution surface [50].

7. Conclusions

The potato is one of the economic crops in Egypt that participates in the providing of local food security and valuable national income through the exportation of agronomic crops. The production of potatoes should be provided with complete care that includes the use of modern safe methods that prevent or treat diseases that affect it, including late blight. Using of effective fungicides is one of these methods which protect from and eradicate the infection of *P. infestans*. The integrated disease management strategy is the best method to control the prevalence of late blight due to the development of new fungicides-resistant strains. The citrus plants are also economic crops in Egypt and require complete care to prevent the growth of the main destructive pathogen called *P. nicotianae* that cause different diseases especially through its association with other pathogens (coinfection).

Acknowledgements

The author sincerely thanks everybody who introduced the required assistance and was helpful to complete this work.

Conflict of interest

The author declares no conflict of interest.

Author details

Waleed Mohamed Hussain Abdulkhair
Egyptian Drug Authority (National Organization for Drug Control and Research), Giza, Egypt

*Address all correspondence to: waleed_hamada@yahoo.com

References

[1] Menge JA, Nemec S. Citrus. In: Hillocks RI, Waller JM, editors. Soilborne Diseases of Tropical Crops. 2nd ed. Wallingford, UK: CAB International; 1997. pp. 185-227

[2] Erwin DC, Ribeiro OK. Phytophthora Diseases Worldwide. 2nd ed. St. Paul, MN, USA: The American Phytopathological Society; 1996. p. 592

[3] Graham JH, Timmer LW, Dewdney MM. Florida citrus pest management guide. In: *Phytophthora* Foot Rot and Root Rot. 2nd ed. USA: Florida University; 2012. pp. 148-156

[4] Nsemwa LTH, Mwalyego FM, Kabungo DA. A review of past crop protection research in the Southern Highlands of Tanzania and prospects for the future. In: Ekpere JA, Rees DJ, Mbwile RP, Lyimo NG, editors. Proceedings of an International Conference on Agricultural Research, Training and Technology Transfer in the Southern Highlands of Tanzania: Past Achievements and Future Prospects. 4th ed. Mbeya, Tanzania: Uyole Agricultural Centre; 1992. p. 1254

[5] Rees DJ, Ekpere JA, Kamasho JA, Gillings SI, ANK M, AEM T. Adapting research to farmers' circumstances in the Southern Highlands of Tanzania. In: Ekpere JA, Rees DJ, Mbwile RP, Lyimo NG, editors. Proceedings of an International Conference on Agricultural Research, Training and Technology Transfer in the Southern Highlands of Tanzania: Past Achievements and Future Prospects. 4th ed. Mbeya, Tanzania: Uyole Agricultural Centre; 1992. p. 1231

[6] Dirac MF, Menge JA, Madore MA. Comparison of seasonal infection of citrus roots by *Phytophthora citrophthora* and *Phytophthora nicotianae* var. *parasitica*. Plant Disease. 2003;**87**:493-501. DOI: 10.1094/PDIS.2003.87.5.493

[7] Jones GD, editor. The Epidemiology of Plant Diseases. 3rd ed. London: Kluwer Academic Publishes; 1998. pp. 371-388. DOI: 10.1007/978-94-017-3302-1

[8] Son SW, Kim HY, Choi GJ, Lim HK, Jang KS, Lee SO, et al. Bikaverin and fusaric acid from *Fusarium oxysporum* show antioomycete activity against *P. infestans*. Journal of Applied Microbiology. 2008;**104**:692-698. DOI: 10.1111/j.1365-2672.2007.03581.x

[9] Latijnhouwers M, Ligterink W, Vleeshouwers VG, VanWest P, Govers F. A Galpha subunit controls zoospore mobility and virulence in the potato late blight pathogen *Phytophthora infestans*. Molecular Microbiology. 2004;**51**:925-936. DOI: 10.1046/j.1365-2958.2003.03893.x

[10] Olanya OM, Adipala E, Hakiza JJ, Kedera JC, Ojiambo P, Mukalazi JM, et al. Epidemiology and population dynamics of *Phytophthora infestans* in Sub-Saharan Africa: Progress and constraints. African Crop Science Journal. 2001;**9**:181-193. DOI: 10.4314/acsj.v9i1.27638

[11] Garrett KA, Dendy SP. Cultural practices in potato late blight management, 107-113. In: Fernandez-Northcoted N, editor. Complementing Resistance to Late Blight (*Phytophthora infestans*) in the Andes. 2nd ed. Bolivia: Cochabamba; 2001. pp. 13-16

[12] Fekede G, Amare A, Nigussie D. Management of late blight (*Phytophthora infestans*) of potato (*Solanum tuberosum*) through potato cultivars and fungicides in Hararghe Highlands, Ethiopia. International Journal Life Sciences. 2013;**2**:130-138

[13] Binyam T, Temam H, Tekalign T. Efficacy of reduced dose of fungicide

sprays in the management of late blight (*Phytophthora infestans*) disease on selected potato (*Solanum tuberosum* L.) varieties Haramaya, Eastern Ethiopia. Journal of Biology, Agriculture and Healthcare. 2014;**4**:46-52

[14] Mesfin T, Gebremedhin WG. Impact of farmers' selected IDM options on potato late blight control and yield. African Crop Science Conference Proceedings. 2007;**8**:2091-2094

[15] Fatma ZR, David ELC, Ivana P, Eva R, Yamina G, Zouaoui B, et al. Characterization of *Phytophthora infestans* populations in northwestern Algeria during 2008-2014. Fungal Biology. 2017;**121**:467-477. DOI: 10.1016/j.funbio.2017.01.004

[16] Binyam T, Temam H, Tekalign T. Tuber yield loss assessment of potato (*Solanum tuberosum* L.) varieties due to late blight (*Phytophthora infestans*) and its management Haramaya, Eastern Ethiopia. Journal of Biology, Agriculture and Healthcare. 2014;**4**:45-54

[17] Fontem DA, Aighew B. Effect of fungicides on late blight control and yield loss of potato in the western highlands of Cameroon. International Journal of Pest Management. 1993;**39**:152-155. DOI: 10.1080/09670879309371781

[18] Shaw DS, Khaki IA. Genetical evidence for diploidy in *Phytophthora*. Genetical Research. 1971;**17**:165-167. DOI: 10.1017/S0016672300012167

[19] Agrios GN, editor. Plant Pathology. 5th ed. London, New York: Academic Press; 2005. p. 922

[20] Fry WE, Goodwin SB, Dyer AT, Matuszak JM, Drenth A, Tooley PW, et al. Historical and recent migrations of *Phytophthora infestans*: Chronology, pathways, and implications. Plant Disease. 1993;**77**:653-661. DOI: 10.1094/PD-77-0653

[21] Injila T, Kabita KS, Subina T, Bindu M, Sudeep S, Jiban S. Late blight of potato and its management through the application of different fungicides and organic amendments: A review. Journal of Agriculture and Natural Resources. 2021;**4**:301-320. DOI: 10.3126/janr.v4i1.33374

[22] Kirk W, Wharton P, Hammerschmidt R, Abu-El Samen F, Douches D. Late Blight. East Lansing, MI: Michigan State University Extension Bulletin E-2945. 2013. Available from: www.lateblight.org

[23] Davis RM, Nunez J, Aegerter BJ. Potato late blight [thesis]. Statewide IPM Program, Agriculture and Natural Resources. University of California; 2009

[24] Ston A. Organic management of late blight of potato and tomato (*Phytophthora infestans*) [thesis]. Sustainable Agriculture Research and Education, Oregon State University. 2009

[25] Anonymous. Knowing your enemy-blight [internet]. 2007. Available from: http://sarvari-trust.org/lateblight.html [Accessed: 5 July 2014]

[26] Beckerman J. Understanding fungicide mobility [internet]. 2008. Available from: Purdue Extension BP-70-W

[27] Hansen EM, Reeser PW, Sutton W. *Phytophthora* beyond agriculture. Annual Review of Phytopathology. 2012;**50**:359-378. DOI: 10.1146/annurev-phyto-081211-172946

[28] Martin FN, Blair JE, Coffey MD. A combined mitochondrial and nuclear multilocus phylogeny of the genus *Phytophthora*. Fungal Genetics and Biology. 2014;**66**:19-32. DOI: 10.1016/j.fgb.2014.02.006

[29] Brasier CM. *Phytophthora* biodiversity: How many Phytophthora

species are there? In: Goheen EM, Frankel SJ, editors. Phytophthoras in Forests and Natural Ecosystems. 2nd ed. Monterey, CA, USA: USDA; 2009. pp. 101-115

[30] Scott P, Burgess TI, Hardy GE. Globalization and *Phytophthora*. In: Lamour K, editor. Phytophthora—A Global Perspective. 3rd ed. Wallingford, UK: CAB International; 2013. pp. 226-232

[31] Meng Y, Zhang Q, Ding W, Shan W. *Phytophthora parasitica*: A model oomycete plant pathogen. Mycology. 2014;**5**:43-51. DOI: 10.1080/21501203.2014.917734

[32] Prigigallo M, Mosca S, Cacciola S, Cooke D, Schena L. Molecular analysis of *Phytophthora* diversity in nursery-grown ornamental and fruit plants. Plant Pathology. 2015;**64**:1308-1319. DOI: 10.1111/ppa.12362

[33] Gallup CA, Sullivan M, Shew HD, Cheryl K. Extension Plant Pathology Support. Black Shank of Tobacco. 2nd ed. USA: Plant Health Instructor; 2006. p. 315

[34] Walker CA, van West P. Zoospore development in the oomycetes. Fungal Biology Reviews. 2007;**21**:10-18. DOI: 10.1016/j.fbr.2007.02.001

[35] Ludowici VA, Zhang W, Blackman LM, Hardham AR. *Phytophthora nicotianae*. In: Lamour K, editor. Phytophthora—A Global Perspective. 4th ed. Wallingford, Oxfordshire, UK: CAB International; 2013. pp. 113-123

[36] Alvarez LA, Gramaje D, Abad-Campus P, GarciaJiménez J. Role of the Helix aspersa snail as a vector of *Phytophthora citrophthora* causing branch cankers on clementine trees in Spain. Plant Pathology. 2009;**58**:956-963. DOI: 10.1111/j.1365-3059.2009.02088.x

[37] Crone M, McComb JA, O'Brien PA, Hardy GE. Survival of *Phytophthora cinnamomi* as oospores, stromata, and thick-walled chlamydospores in roots of symptomatic and asymptomatic annual and herbaceous perennial plant species. Fungal Biology. 2013;**117**:112-123. DOI: 10.1016/j.funbio.2012.12.004

[38] Cacciola SO, Magnano G. Management of citrus diseases caused by *Phytophthora* spp. In: Ciancio A, Mukerji KG, editors. Integrated Management of Diseases Caused by Fungi, Phytoplasma and Bacteria. 3rd ed. Dordrecht, The Netherlands: Springer Science; 2008. pp. 61-84

[39] Drenth A, Guest DI, editors. Diversity and Management of *Phytophthora* in Southern Asia. 3rd ed. Australian Centre for International Agricultural Research Canberra (ACIAR) Monograph 114; 2004. p. 238

[40] Alvarez LA, León M, Abad-Campus P, García-Jiménez J, Vicent A. Genetic variation and host specificity of *Phytophthora citrophthora* isolates causing branch cankers in Clementine trees in Spain. European Journal of Plant Pathology. 2011;**129**:103-117. DOI: 10.1007/s10658-010-9696-8

[41] Vernière C, Cohen S, Raffanel B, Dubois A, Venard P, Panabières F. Variability in pathogenicity among *Phytophthora* spp. pathogenic to citrus in corsica. Journal of Phytopathology. 2004;**152**:476-483. DOI: 10.1111/j.1439-0434.2004.00878.x

[42] Graham JH, Bright DB, McCoy C. *Phytophthora—Diaprepes* weevil complex: *Phytophthora* spp. relationship with Citrus rootstocks. Plant Disease. 2003;**87**:85-90. DOI: 10.1094/PDIS.2003.87.1.85

[43] Wang N, Trivedi P. Citrus Huanglongbing: A new relevant disease presents unprecedented challenges.

Phytopathology. 2013;**103**:652-665. DOI: 10.1094/PHYTO-12-12-0331-RVW

[44] Gottwald TR. Current epidemiological understanding of citrus Huanglongbing. Annual Review of Phytopathology. 2010;**48**:119-139. DOI: 10.1146/annurev-phyto-073009-114418

[45] Graham JH, Johnson EG, Gottwald TR, Irey MS. Presymptomatic fibrous root decline in citrus trees caused by Huanglongbing and potential interaction with *Phytophthora* spp. Plant Disease. 2013;**97**:1195-1199. DOI: 10.1094/PDIS-01-13-0024-RE

[46] Graham JH, Feichtenberger E. Citrus Phytophthora diseases: Management challenges and successes. Journal of Citrus Pathology. 2015;**2**:1-11. DOI: 10.5070/C421027203

[47] Duncan LW, Rogers ME, Futch SH, Graham JH. 2014 Production guide: Citrus root weevils [thesis]. UF/IFAS Extension, Florida Citrus Production Guide, University of Florida; 2014

[48] Caten CE, Jinks JL. Spontaneous variability of single isolates of *Phytophthora infestans*. I. Cultural variation. Canadian Journal of Botany. 1968;**46**:329-348. DOI: 10.1139/b68-055

[49] Hartman GL, Huang YH. Characteristics of *Phytophthora infestans* isolates and development of late blight on tomato in Taiwan. Plant Disease. 1995;**79**:849-852

[50] Andrivon D, Avendano-Corcoles J, Cameron AM, Carnegie SF, Cooke LR, Corbiere R, et al. Stability and variability of virulence of *Phytophthora infestans* assessed in a ring test across European laboratories. Plant Pathology. 2001;**60**:556-565. DOI: 10.1111/j.1365-3059.2010.02392.x

Chapter 2

Phytophthora Diseases in Ghana: Threats, Management Options and Prospects

Joseph Adomako, Stephen Larbi-Koranteng, Frederick Kankam, Yaw Danso, Jerry Fenteng Asamoah, Patricia Oteng-Darko, Kennedy Agyeman, Stephen Yeboah, Erasmus Narteh Tetteh and Eric Owusu Danquah

Abstract

Ghana's agricultural economy is largely dominated by the crop subsector with much focus on the production of tree, arable and vegetable crops. Nevertheless, *Phytophthora* spp. are major threat to the production of these crops contributing significantly to yield reduction. In this review, the main focus will be to look at the threats the pathogen poses to production, economic importance of *Phytophthora* diseases, highlights some *Phytophthora* diseases with limited research in the country but have the potential of affecting crop production, management options and the prospect of developing and deploying biological control strategies considered environmentally friendlier and devoid of human health risks to reduce the effect of this pathogen on crop production as well as reducing the dependency on chemical control option.

Keywords: Biocontrol, Chemical control, Epidemiology, *Phytophthora*, Symptoms

1. Introduction

Plant diseases are of significant importance in plant production globally. From the resource poor small-scale farmer to the most sophisticated commercial grower, plant diseases remain a major source of concern affecting all levels of the farming enterprise contributing significantly towards achieving a reliable world's food systems. It influences decision making process of not only producers but traders, processors and even consumers contributing immensely towards the various components of food systems. Directly, diseases impact negatively on yield and it is estimated that about 16% of global arable food is lost to diseases [1, 2]. Several pathogens cause plant diseases, however, since the infamous Irish potato famine in 1840, the pathogen, *Phytophthora* sp., has been considered the most destructive plant pathogen [3] with more than 80 species in the genus with multiple or diverse host ranges [4, 5]. Several documents have reported on the negative impact of *Phytophthora* diseases on global economy in relation to human suffering, hunger, disease and subsequent death [6, 7]. Yield losses associated with the pathogen is estimated in billions of dollars annually [8, 9].

For years, *Phytophthora* diseases have impacted tremendously on Ghana's agriculture threatening the economic livelihood of farmers most especially those in cocoa production sector. *Phytophthora* pod rot disease commonly known as the Black pod disease (**Figure 1**) caused by *Phytophthora palmivora* and *P. megakarya* is the most destructive cocoa disease worldwide accounting for over 30% pod yield loss through pod rot and 10% tree death [11]. Since the incidence of *Phytophthora* diseases of cocoa were reported in Ghana [12], high yield loss ranging between 60 and 100% has been recorded [12] with the country losing more than a quarter of its 2012 annual output of 850,000MT to *Phytophthora* spp. disease [13]. Continuous destruction of cocoa by *Phytophthora* has compelled most farmers to abandon cocoa production for other crops [14, 15].

Although *Phytophthora* diseases on cocoa have received a lot of attention and well documented in Ghana, reports on their impact on other crops are scanty and highly unavailable except for taro [*Colocasiae esculenta*] which recently has been documented. Since the initial report of taro leaf blight [16] in the Eastern region of Ghana, several researchers [17, 18] have reported the prevalence of *Phytophthora* leaf blight of taro (**Figure 2**) and morphological and genetic variations in isolates of the pathogen from different parts of the country [19]. Apart from reduction in yield, the pathogen is reported to have caused most farmers to shift from taro production to other crops such as sugar cane and rice [17]. Akrofi et al. [20] reported that plants such as *Xanthosoma saggitifolium*, *Musa paradisiaca*, *Carica papaya*, *Ananas comosus*, *Elaeis guinnensis*, *Persia americana* and *Mangifera indica* commonly used as shade plants in cocoa production, served as alternative hosts to *Phytophthora*. This is worrying as little research has so far been geared towards studying the impact of *Phytophthora* on several crops in Ghana. This clearly poses a threat to crop production especially as several reports reveal susceptibility of most of the country's food and cash crops to different *Phytophthora* species leading to massive decline in their production. For example, diseases such as the tomato late blight, pineapple heart rot, and root rot of papaya which are caused by *Phytophthora* species can greatly cause fruit and vegetable insecurity in the event of future outbreak when knowledge on them is low. These challenges and gaps present opportunities for research into *Phytophthora* diseases in the country such as assessing population structure and determining the epidemiology of *Phytophthora* diseases, identifying host ranges and designing management strategies to minimize its impact. Impact studies of the pathogen can also be carried out to quantify their effect on crop production. The aim of this review is not to exhaust all aspects on *Phytophthora* diseases in Ghana but bring to light current and potential threats on some important crops, control options being used by farmers and the need to research and promote the use of biocontrol agents in Ghana. It is believed this review will inspire Plant Pathologists to delve deep into *Phytophthora* research in Ghana.

Figure 1.
Pods showing symptoms of black pod diseases [10].

Figure 2.
Symptoms of taro leaf blight [17].

2. *Phytophthora* diseases of economic importance

Characteristics of all host plant impairments, *Phytophthora* diseases generally interrupt with the normal physiological functions of their host thereby reducing productivity and consequently lead to food and economic insecurity among populations which depend on it. Globally, impact of the late blight disease of potato, caused by the *P. infestans* [Mont.] de Bary showed clearly the significance of plant diseases and more especially those caused by *Phytophthora* species. This epidemic led to mass starvation, death and migration of people from Ireland to the United States [21]. In Ghana, the impact of *Phytophthora* diseases is highly prominent and its impact heavily felt in cocoa production. The crop is a major earner for the country accounting for about 67% of household income for about 25–30% of Ghanaians living across the cocoa growing regions [22]. As important as the economic value of cocoa to the country, *Phytophthora palmivora* and *P. megakarya* infections has been reported to cause a stagnating effect on its production. The disease which was first reported in Ghana in 1985 [12], covering an estimated area of about 16, 000 hectares of cocoa farm land is now prevalent across all cocoa growing areas in the country [10]. The black pod disease, caused by *Phytophthora* sp. has been described as the single most destructive limitation to the economic production of cocoa. Apart from the heavy pod loss, *Phytophthora* spp. causes stem canker in cocoa leading to the death of plant [23]. This clearly shows a reduction in plant population and consequently yield of the farm. Omane et al. [16] identified *Phytopthora colocasiae* as causing taro leaf blight disease in the country apart from *Phytophthora palmivora* and *P. megakarya* on cocoa. The disease has been associated with about 90 and 50% leaf and corm yield losses respectively. High incidence and severity of the disease was reported in eleven districts of the semi-deciduous agro ecologies of Ghana [17]. It causes corm rot and invasion of the rot by other pathogens such as *Lasiodiplodia theobromae* causing the corms to blacken in storage [24]. The major characteristic feature of all countries where the disease had been reported is the forced abandonment of taro production by farmers or replanting of fields with crops like sugar cane and rice [17]. This increases food insecurity in taro growing communities as the crop is used as a substitute to major food crops during lean seasons. A survey by [25] revealed high incidence of citrus trunk rot in plantations in Ghana. Reports later showed *Phytophthora citrophthora* as major cause of the disease as it causes citrus gummosis, [26] affecting major *Citrus* species such as sweet orange [*C. sinensis*], lemon [*C. limon*], mandarin [*C. reticulata*], grapefruit [*C. paradisi*], and lime [*C. aurantifolia*] [27]. The pathogen is reported to cause tree cankers leading to death and decline of the tree crop in the field. Despite several interventions by farmers and various stakeholders to possibly eliminate the menace of this disease, it had not been successful which buttressed its economic importance in citrus production [28].

3. Neglected *Phytophthora* diseases on crops of economic importance

Majority of research reports on *Phytophthora* diseases, in Ghana have focused on cocoa, with limited number on taro and citrus and virtually none existing for other crops. The pathogen, however, has been reported to be a great limitation to the production of many crops produced globally. Various species of *Phytophthora* cause varying degrees of damage to fruits, legumes, orchards and vegetable crops. The pathogen is responsible for several plant diseases such as late blight, root, stem and fruit rots in several crop species from different families such as Bromeliaceae [*Ananas comosus*, Pineapple], Caricaceae [*Carica papaya*- Pawpaw], Sterculiaceae [*Cola nitida*-bitter cola and *Sterculia tragacantha*], Agavaceae [*Dracaena manni*], Arecaceae [*Elaeis guinneensis*-African oil palm], Apocynaceae [*Funtumia elastic* -West African Rubber tree], Anacardiceae [*Magnifera indica*-Mango], Musaceae [*Musa x paradisiaca*], Lauraceae [*Persia americana*-Avocado], Euphobiaceae [*Ricinodendron heudelotii*-njanysa] and *Inviigia* sp. [20], although reports on disease assessment, prevalence and economic importance have not been established in Ghana. In several countries where *Phytophthora* diseases have been assessed and reported in any crop species, pre and post-harvest losses due to even latent infections during transportation and storage has been huge. Akrofi et al. [20] reported the role of some economic crops on cocoa farms in black pod disease epidemiology. Since these economic plants serves as alternative host to *Phytopthora* species on cocoa farms, then the potential of a possible disease outbreak due to *Phytophthora* cannot be ignored. Fresh fruits and vegetable crops produced in Ghana are exported to several countries for foreign exchange. *Phytophthora* infections can therefore lead to the rejection of these produce at various quarantine checks at points of entry. Furthermore, any delay in the transport of infected fruits or other produce to their destination of use will lead to huge loss as disease development is very fast when conditions are favorable.

It is worth noting that *Phytophthora* disease does not only have impact on the produce but also on the general cost of production. With the objective to manage the severity of the infestation of the pathogens, most farmers employ several methods among which chemical method of control is most preferred due to the quick result it provides. The cost of acquiring these chemicals in the long run increases the production cost of the farmer. Chemicals used in controlling most of these diseases in most cases turn to react with the environment. This leads to the contamination of water bodies and destruction of other useful microorganisms in the ecosystem. *Phytophthora* diseases have the potential of threatening the agricultural economy of the country with the potential to cause food insecurity. Therefore, serious attention needs to be paid to the disease to limit the menace of its impact on the agriculture enterprise.

4. Epidemiology and symptoms of *Phytophthora* diseases

Impact of climate conditions on the incidence and severity of the *Phytophthora* spp. have been reported [29]. Ndoumbe'-Nkeng [30] postulated that high levels rainfall, higher relative humidity, and lower atmospheric temperatures are known to be the main conditions favorable for the development of Phytophhtora diseases. In agreement to this, Deberdt et al. [29] also observed a significant positive correlation between rainfall and incidence of the *P. megakarya* pod rot. It has been observed in Ghana that the occurrence of the disease is between July and October across the cocoa growing areas [15, 31]. According to [32], Phythophtora disease cycle is denoted by a parasitic phase which occurs during wet and dry seasons.

Phytophthora spp. over-seasons on pod husks, in the soil, leaf debris and or roots or shade plants [33]. When conditions are favorable [mostly during the rainy season], there is a germination of the fungi sporangia, followed by the releasing of motile zoospores of the *Phythophtora* mostly in free water and quickly spread to cause destruction of host plant. *Phytophthora megakarya* and *P. palmivora* for example undergoes series of developmental phases to form the disease cycle (**Figure 3**). Symptoms of *Phytophthora* infections can be observed on every part of an infected plant mostly under wet or humid conditions. In almost all foliar diseases resulting from *Phytophthora* infection, the appearance of small translucent spot are the initial symptoms which later appears as brown to darkened spots of lesions on the affected parts [11]. These spots coalesce as environmental conditions favors it. In cocoa for example, infected pods result in browning, blackening, shriveling up, or total rotting of the pod [20]. Infected roots cause plants to pull up easily from the soil due to root

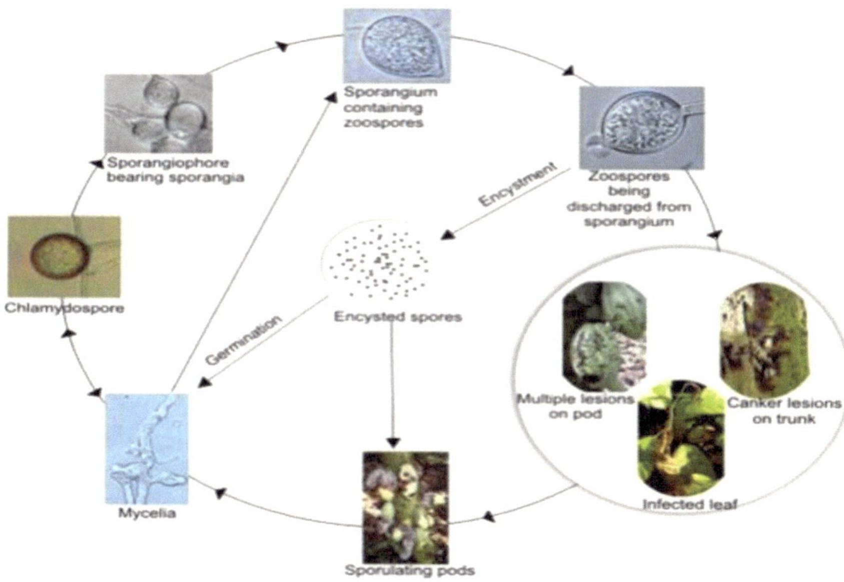

Figure 3.
Disease cycle of P. megakarya on cocoa [20].

Figure 4.
Stem canker of cocoa [www.pestnet.org].

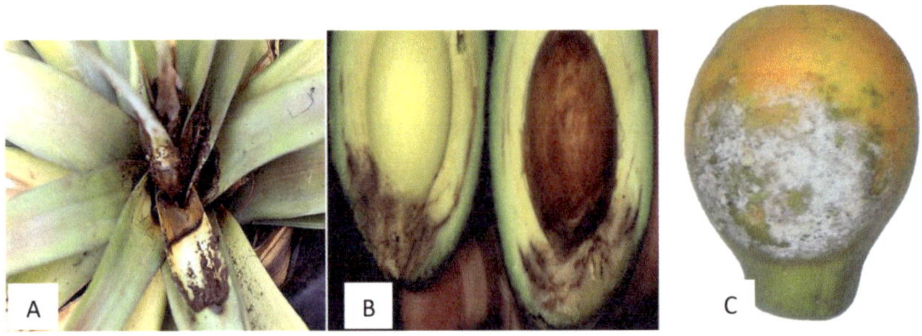

Figure 5.
Symptoms of Phytophthora *infection causing pineapple heart rot (A), Avocado fruit rot (B) and Papaya fruit rot (C). [34, 35] (https://www2.ipm.ucanr.edu/agriculture/avocado/Phytophthora-fruit-rot/).*

rot and loss. In most instances, mycelial growth covers infected plant parts areas under moist conditions. *Phytophthora* diseases can also cause cankers (**Figure 4**), fruit rots (**Figure 5**) and leaf blight (**Figure 2**).

5. Management of *Phytophthora* diseases

Crop disease management is one of the heaviest cost burdens in the crop production enterprise. The neglect of diseased farms without control or management interventions can cause a serious loss to the farmer. Management of plant diseases aims at either attacking the pathogen directly or creating a condition that will be unfavorable for its establishment and development. Achieving this calls for the employment of several possible control methods and strategies.

For years, various fungicides have been employed in the control of *Phytophthora* diseases most especially black pod disease in cocoa [36–39] and recently *Phytophthora* leaf blight of taro in the country [40, 41]. The chemical control relies mainly on copper-based fungicides and systemic fungicides such as metalaxyl and phosphonates. Among them are Kocides 101 [77% cupric hydroxides], Cocoabre Sandox [56% cuprious oxides], Copper Nordox 50 [50% cuprous oxide], Copper Nordox 75 [Copper oxide] Champion [77% cupric hydroxide], Ridomil 72 plus [12% metalaxyl and 60% cuprous oxide] among others for the control of *P. palmivora* and *P. megakarya*, the two main causal agents of *Phytophthora* pod rot of cocoa [42]. The effectiveness of fungicide for the control of *Phytophthora* disease, however, depends on factors such as the climatic conditions at the time and location of application, the crop variety, and pathogen species among others like social and economic considerations [38]. To be able to achieve an effective chemical control of *P. megakarya*, fungicides have to be sprayed at shorter intervals. In agreement with this, [15] reported that whiles a by-weekly application was recommended in Cameroon due to high and frequent rainfall, an average of 4-weeks intervals in Ghana was effective. This deference could be attributed to the differences in the frequency and amount of rainfall in Cameroon and Ghana. Chemical control in most cases have been reported to be costly. This makes chemical control unattractive to many peasant farmers. However, this approach could be cost effective when crop price is comparatively high and the crop is also under a low disease pressure [14].

Cultural control strategies according to [43], is one of the ancient strategies in the management of plant diseases. It is the strategic use of the day-to-day farm practices to either inhibit, obstruct the establishment, growth and development of the pathogen. Cultural control system has proved not only to increase yield but also created the conducive environment for efficient performance of applied fungicides [44]. To the low-income farmers, cultural control is the most cost-effective disease management approach. This is because it does not require any extra cost apart from that which has already been the situation. For instance, early harvesting, removal of infected pods and pruning of infected branches reduces the inoculum load present by the primary or secondary infection. As postulated by [32], the appropriate tree spacing improves aeration, reduces huge canopy humidity and also keeps leaf mulch or litter in check. However, Luterbacher [45] in contrast of this report also reported that leaf litter has no major impact in reducing cocoa pod infection from soil inoculum. Despite the promising impact of this method, it was also observed that the sole implementation of this approach could be labour intensive and thus needed to be complemented with other control methods [30, 31, 46].

With frequent travel and trade within Ghanaian cities and regions, the fast spread of *Phytophthora megakarya* has been linked to the movement of plant materials from one district to another [15, 23]. This poses a high risk as several plant diseases are moved from infected regions to an uninfected region. Quarantine method could be adopted to overcome this menace. The quarantine system of plant disease control ensures that agriculture and natural resources in a localized area or district or region are safeguarded against the entry of disease pathogens. This will help to ensure an abundant, high-quality, and varied food supply within the territory and beyond.

6. Novel approaches to control *Phytophthora* diseases in Ghana: success, challenges and prospects

In the management of *Phytophthora* diseases in Ghana, especially cocoa diseases, much emphasis has been laid on chemical and cultural control as against other control options such as biological control. In the chemical control of *Phytophthora* disease, there has not been only an over reliance but an abuse of synthetic fungicides which have both health and environmental effects although the strategy is quicker, reliable and effective [47]. Utilization of control option such as biological control is not only imperative but possible, since there are numerous reported cases of biological control attempts [48–51]. Biological control strategy if harnessed will not only become a complementary but also an alternative for the control of *Phytophthora* disease as it is considered as an environmentally friendly form of disease control [52]. Biological control is the judicious use of an antagonist, its parts or product (antibiotic/secondary metabolite) to inhibit, prevent or control plant diseases. Directly, it protects their host using mechanisms such as competition for nutrients, root colonization and competition for infection sites, secretion of extracellular lytic enzymes and hyperparasitism, and induction of plant resistance. Indirectly, it promotes plant growth thereby protecting the plant from pathogen's attack [53, 54]. Not much reported success stories on the use of biocontrol agents against *Phytophthora* diseases in Ghana, but several attempts such as identifying microbial antagonists, botanicals, and resistant varieties have been initiated which needs to be appreciated.

There have been reports on the use of natural agents against *P. palmivora* and *P. megakarya* in Ghana especially *in vitro*. Not much work have been done in the field. The first of such reports was by Attafuah [55] who demonstrated that there was an inhibition of *P. palmivora* by an isolate of *Pseudomonas aeruginosa* [Schröter] Migula on cocoa husk *in vitro*. It was reported that the antagonist was isolated from cocoa mealybug, *Planococcoides njalensis* Laing and tested against *P. palmivora*. It was through the effort and the works of Attafuah that formed the basis for Odamtten and Clerk [56] to work with *Aspergillus niger* and *Trichoderma viride*. Their results showed that metabolites of *A. niger* and *T. viride* inhibited zoospore motility, direct germination and indirect germination of sporangia, mycelial growth, sporulation and sporangial size of *P. palmivora in vitro*. Akrasi [57] isolated and identified eight *Bacillus* species strains from yam rhizosphere which were antagonistic to *P. palmivora*, the causal agent of cocoa pod rot disease (Black pod disease as being referred to in Ghana). Akrasi [57] elucidated that the filtrates of the rhizobacteria were fungitoxic and thermostable when exposed to temperature of about 121°C during autoclaving. He also reported that the filtrates of the bacteria were comparable in their effect to two fungicides, Thiophanate Methyl [Topsin M 70 WP] and Ridomil 72 plus [72% WP].

Based on the findings of Akrasi [57], [58] applied the promising rhizobacteria as protectant on cocoa pods and reported that that both broth culture and culture filtrate of the rhizobacterium isolates applied, completely inhibited *P. palmivora* infection and lesion development on detached cocoa pods. In recent studies, the rhizobacterium was identified as *Bacillus amyloliquefaciens* and field studies conducted alongside with two other antagonist viz. *Aspergillus* and *Penicillium* spp. showed these three antagonists as having potential to be developed as biocontrol agents against the black pod disease of cocoa [59].

Apart from the numerous inorganic pesticides, reports on the use of plant-based fungicides against *Phytophthora* in Ghana is limited. However, Awuah [60] postulated that natural substances from plants possess antimicrobial effects that could be potentially used in the control of *Phytophthora* diseases. This was based on the fact that, crude steam distillate of *Ocimum gratissimum* completely inhibited the growth of the pathogen and prevented black pod development on detached cocoa pods. The author further reported that the use of *Cymbopogon citratus* and *O. gratissimum* against the black pod pathogen in the field were effective against the black pod pathogen comparable to Kocide 101 [60].

For years several attempts have been made in the identification, breeding and selection of germplasm especially in cocoa for resistance to *P. palmivora* and *P. megakarya*. The use of such traditional breeding techniques for the breeding of cocoa resistance to *P. palmivora* and *P. megakarya* has been of little success. There is a well-documented reports on screening of cocoa accessions for resistance to *P. palmivora* [61]. There was also a programme to assess the existing cocoa germplasm in Ghana for resistance or tolerance to *P. megakarya* [62]. However, [63] reported that there were no cocoa genotypes with immunity to the black pod disease pathogens in Ghana. Modern breeding approaches are currently employed to determinine the mode of inheritance, combining ability and heritability of resistance to *P. palmivora* and *P. megakarya* in cocoa germplasm [64]. Cocoa hybrids such as Alpha B36xPa7/808, Pa7/808 pound 7 and Alpha B36 xT65/326 were identified to possess different levels of resistance against major cocoa diseases [65]. With respect to taro, some germplasm have shown high level of tolerance to *P. colocasie* [19, 66].

The main challenges confronting the development and use of biological control as complementary/alternative to chemical control options in Ghana are the lack of Government initiative to promote biological control as against the importation and use of synthetic fungicides that have health and environmental effect. In some countries, there are policies to promote the use of biological control agents for the control of plant diseases. Through the National Research Initiative and other USDA programmes, research funds are made available for funding biological control activities. Among such funds are the Section 406 programme, IR-4, Regional IPM grants, and Integrated Organic Programme. Funds are also made available to stimulate the development of commercial ventures for the small business innovative research [SBIR] programmes [67]. The Government of Ghana therefore has to make it as a policy to promote research and encourage the usage of novel control options to complement the use of chemicals in an integrated manner. Scientific research in

Ghana on biological control needs to be coordinated leading to the establishment of Biological Control Community of Practice to encourage effective research and promotion of biocontrol agents. Farmers and other stakeholders need continuous education to raise their awareness on the dangers of abusing synthetic pesticides and its effect on health and the environment.

Notwithstanding the challenges, the future looks bright for the development of biological control and other control strategies in the country to mitigate the effects of *Phytophthora* due to the fact that awareness on the impact of the pathogen on crop production is gaining attention and research to characterized the structures and functions of biological control agents, pathogens, and host plants at molecular, cellular, organismal, and ecological levels are gradually receiving attention of plant pathologists and breeders in the country. To ensure that world food becomes safer in the next decade, then there should be increase in demand for safer pesticide in Agriculture and the solution should therefore be a biological control in an integrated management [IM] systems.

7. Conclusion

Phytophthora disease and its impact as a potential cause of food insecurity is something that every economy has to seriously take into consideration. It affects a wide range of crops; fruits, vegetables, legumes, root and tubers. It causes total yield loss, both in the field, transport or even in storage due to latent infection. Despite the threat posed by this pathogen, limited studies have been carried out on most crops in Ghana. In this chapter, we able to elucidate the various crops that are affected by the pathogen, the effort being made to manage it, the prospect of using biocontrol agents against the use of chemical fungicides due to their effect on the environment and humans. The review has also thrown more light on how concentration has been on *Phytophthora* pod rot disease of cocoa at the neglect of other equally important diseases caused by the pathogen. Highlighting the challenges confronting the use biocontrol as an option, the main issue addressed was lack of governmental policy to promote this practice and insufficient funding to promote research on *Phytophthora*. More attention is therefore needed with the aim of limiting its impact on food and tree crops production in Ghana.

Conflict of interest

Authors declare no competing interest.

Author details

Joseph Adomako[1*], Stephen Larbi-Koranteng[2], Frederick Kankam[3], Yaw Danso[1], Jerry Fenteng Asamoah[1], Patricia Oteng-Darko[1], Kennedy Agyeman[1], Stephen Yeboah[1], Erasmus Narteh Tetteh[1,4] and Eric Owusu Danquah[1]

1 CSIR-Crops Research Institute, Kumasi, Ghana

2 College of Agricultural Education, Apenten Appiah-Menka University of Skills Training and Entrepreneurial Development, Asante-Mampong, Ghana

3 Department of Crop Science, Faculty of Agriculture, Food and Consumer Sciences, University for Development Studies, Tamale, Ghana

4 Department of Agroforestry, Kwame Nkrumah University of Science and Technology, Kumasi, Ghana

*Address all correspondence to: joeadomako@gmail.com

References

[1] Oerke EC. Crop losses to pests. The Journal of Agricultural Science, 2006; 144[1], 31-43.

[2] Savary S, Bregaglio S, Willocquet L, Gustafson D, Mason D'Croz D, et al. 2017. Crop health and its global impacts on the components of food security, Food Secururity, 2017; 9:311-327.

[3] O'Brien PA, Williams N, Hardy GES. Detecting Phytophthora. Critical Reviews in Microbiology, 2009; 35[3], 169-181.

[4] Blair JE, Coffey MD, Park SY, Geiser DM, Kang S. A multi-locus phylogeny for Phytophthora utilizing markers derived from complete genome sequences. Fungal Genetics and Biology, 2008; 45[3], 266-277.

[5] Erwin DC, Ribeiro OK. Phytophthora diseases worldwide. American Phyto-pathological Society [APS Press], 1996.

[6] Kamoun S, Furzer O, Jones JD, Judelson HS, Ali GS, Dalio, RJ, Govers F. The Top 10 oomycete pathogens in molecular plant pathology. Molecular Plant Pathology, 2015; 16[4], 413-434.

[7] Yadav S, Lal M, Singh BP, Kaushik SK, Sharma S. Evaluation of fungicides against *Phytophthora infestans* at elevated temperature. Potato Journal, 2016; 43[1].

[8] Judelson HS, Blanco FA. The spores of Phytophthora: weapons of the plant destroyer. Nature Reviews Microbiology, 2005; 3[1], 47-58.

[9] Latijnhouwers M, Ligterink W, Vleeshouwers VG, van West P, Govers F. A Gα subunit controls zoospore motility and virulence in the potato late blight pathogen *Phytophthora infestans*. Molecular microbiology, 2004; 51[4], 925-936.

[10] Akrofi AY, Assuah MK, Amoako Atta I. Procedure for screening "new fungicides" for the control of Phytophthora pod rot [black pod] disease of cocoa in Ghana. Cocoa Research Institute of Ghana, 2013; 27, 1-14.

[11] Guest D. Black Pod: Diverse pathogens with a global impact on cocoa yield. Phytopathology, 2007; 97[12]: 1650-1653.

[12] Dakwa JT. A serious outbreak of blackpod disease in a marginal area of Ghana. Proceedings of the 10th International Cocoa Research Conference, Santo Domingo, Dominican Republic, 17- 23rd May, 1987.

[13] Ghana Cocobod. Cocobod Annual Report, 2012; https://cocobod.gh/resources/annual-report Accessed June 2, 2021

[14] Akrofi AY, Appiah AA, Opoku IY. Management of Phytophthora pod rot disease on cocoa farms in Ghana. Crop Protection, 2003; 22:469-477.

[15] Opoku IY, Appiah AA, Akrofi AY, Owusu GK. *Phytophthora megakarya* potential threat to the cocoa industry in Ghana. Ghana Journal of Agricultural Science, 2000; 33:237-248.

[16] Omane E, Oduro KA, Cornelius EW, Opoku IY, Akrofi, AY, Sharma K, Bandyopadhyay, R. First report of leaf blight of taro [*Colocasia esculenta*] caused by *Phytophthora colocasiae* in Ghana. Plant disease, 2012; 96[2], 292-292.

[17] Adomako J, Kwoseh CK, Moses E, Larbi-Koranteng S. Prevalence of Phytophthora Leaf Blight of Taro [*Colocasia esculenta* [L.] Schott] in the Semi Deciduous Forest Zone of Ghana. Journal of Experimental Agriculture International, 2016; 1-7.

[18] Van der Puije GC, Ackah FK, Moses E. Prevalence of leaf blight

disease caused by *Phytophthora colocasiae* in taro in the Aowin Suaman Districts of Ghana. Hort Flora Research Spectrum, 2015; 4 [3]: 282, 284.

[19] Adomako J, Kwoseh C, Moses E, Prempeh RNA. Variations in morphological and molecular characteristics of *Phytophthora colocasiae* population causing leaf blight of taro in Ghana. Archives of Phytopathology and Plant Protection, 2019; 51[19-20]:1009-1021.

[20] Akrofi AY, Amoako-Atta I, Assuah M, Asare EK. Black pod disease on cocoa [Theobroma cocoa, L] in Ghana: spread of *Phytophthora megakarya* and role of economic plants in the disease epidemiology. Crop Protection, 2015; 72, 66-75.

[21] Achouri A. From Ireland to America: Emigration and the Great Famine 1845 – 1852, International Journal of Humanities and Cultural Studies, 2016; 2[4]:22-38.

[22] Kolavalli S, Vigneri, M. Cocoa in Ghana: Shaping the success of an economy. Yes, Africa can: success stories from a dynamic continent, 2011; 201-218.

[23] Opoku IY, Akrofi AY, Appiah AA. The spread of *Phytophthora megakarya* on cocoa in Ghana. Proceedings of the 1st International Cocoa Pests and Diseases Seminar, Accra, Ghana, 6-10th November, 1997.

[24] Singh R, Kalra A, Ravish BS, Divya S, Parameswaran TN, Srinivas KVNS, Bagyaraj DJ. Effect of potential bioinoculants and organic manures on root-rot and wilt, growth, yield and quality of organically grown *Coleus forskohlii* in a semiarid tropical region of Bangalore [India], Plant pathology, 2012; 61[4], 700-708.

[25] Leather RI. Diseases of economic plants in Ghana other than cocoa. Bulletin of Ministry of Food and Agriculture, 1959; 1:1-40.

[26] Brentu FC, Vicent A. Gummosis of citrus in Ghana caused by *Phytophthora citrophthora*. Australasian Plant Disease Notes, 2015; 10[1], 1-3.

[27] Assuah MK, Oduro KA, Ofuso-Budu KG. *Diplodia natalensis* Pole Evans, the causal agent of citrus gummosis disease in Ghana. Ghana Journal of Agricultural Science, 1999; 32[1], 11-18.

[28] Ofosu-Budu KG, Monney EO, Quaye E, Amankwah A, Mintah P, Mpere-Asare C, Agboka M. Citrus production in Ghana. Horticulture Export Industry Initiative, Ministry of Food and Agriculture, Accra, 2007.

[29] Deberdt P, Mfegue CV, Tondje PR, Bon MC, Ducamp M, Hurard C. Impact of environmental factors, chemical fungicide and biological control on cacao pod production dynamics and black pod disease (Phytophthora megakarya) in Cameroon. Biological Control, 2008; 44:149-159.

[30] Ndoumbe-Nkeng, M, Cilas, C, Nyemb, E, Nyasse, S, Bieysse, D, Flori, A. Impact of removing diseased pods on cocoa black pod caused by *Phytophthora megakarya* and on cocoa production in Cameroon. Crop Protection, 2004; 23:415-424.

[31] Opoku IY, Akrofi AY, Appiah AA. Assessment of sanitation and fungicide application directed at cocoa tree trunks for the control of Phytophthora black pod infections in pods growing in the canopy. European Journal of Plant Pathology, 2007a; 117:167-175.

[32] Gregory PH, Griffin MJ, Madddison AC, Ward MR. Cocoa black pod: A reinterpretation. Cocoa Growers' Bulletin, 1984; 35: 5-22.

[33] Opoku IY, Akrofi AY, Appiah AA., Shade trees are alternative hosts of the cocoa pathogen Phytophthora megakarya. Crop Protection, 2002;

21[8], 629-634. doi: 10.1016/S0261-2194[02]00013-3

[34] Green J, Nelson S. Heart and root rots of pineapple. College of Tropical Agriculture and Human Resources, University of Hawaii, Manoa, Honololu, 2015.

[35] Liberato JR, Suzuki MS, Denth A. Phytophthora Fruit Rot and Root Rot of Papaya (Phytophthora palmivora), 2006; Available online: PaDIL - http://www.padil.gov.au.

[36] Bateman R, Hidalgo E, García J, Hoopen GT, Krauss U, Adonijah V, Catie T. Rational fungicide use in cocoa: Improving agents and application techniques. Semana Científica 2004. Memoria. 6. Semana Científica. Turrialba [Costa Rica]. 11-12 Mar 2004.

[37] Hidalgo E, Bateman R, Krauss U, Hoopen M, Martínez A. A field investigation into delivery systems for agents to control *Moniliophthora roreri*. European Journal of Plant Pathology, 2003; 109: 953-961.

[38] Norgrove, L. Effects of different copper fungicide application rates upon earthworm activity and impacts on cocoa yield over four years. European Journal of Soil Biology, 2007; 43, S303-S310.

[39] Russell PE. A century of fungicide evolution. Journal of Agricultural Science,2005; 143: 11-25.

[40] Adomako J, Amengor NE, Larbi-Koranteng S, Kankam F. Efficacy of spraying intervals of Ridomil Plus 66 WP for control of taro leaf blight disease. African Crop Science Journal, 2021; 29[1], 133-140.

[41] Omane E, Oduro KA, Cornelius EW, Akrofi AY, Asare EK, Barnabas AA. Improvement in the Management of Leaf Blight Disease of Taro [*Colocasia esculenta*] with Copper and Metalaxyl Fungicides. Journal of Experimental Agriculture International, 2020; 127-136.

[42] Akrofi AY, Opoku IY, Appiah AA. [1995]. On-farm farmer managed trials to control black pod disease caused by *Phytophthora megakarya* in Ghana. In Proceedings of the First International Cocoa Pests and Disease Seminar, Accra, Ghana [pp. 6-10].

[43] Sitapai EC. Studies on resistance in cocoa [*Theobromae cocoa* L.] to *Phytophthora palmivora* [Butler], Doctoral dissertation, MSc thesis, submitted to University of West Indies, 1989; 234p.

[44] Akrofi AY, Opoku IY, Appiah AA. On farm farmer-managed trials to control black pod disease caused by *Phytophthora megakarya* in Ghana. Proceedings First International Cocoa Pests and Diseases Seminar, Accra, Ghana, 1997; 6-10 November.

[45] Luterbacher, MC. The identification, epidemiology and control of *Phytophthora megakarya* on cocoa in West Africa, 1994; Doctoral dissertation, University of London.

[46] Opoku IY, Assuah MK, Aneani F. Management of black pod disease of cocoa with reduced number of fungicide application and crop sanitation. African Journal of Agricultural Research, 2007b; 2:601-604

[47] Adejumo TO. 1997. Identification, incidence, severity and methods of control of the causal organism of false smut disease of cowpea, *Vigna unguiculata* [L.] Walp [Doctoral dissertation, PhD. Thesis. 1997; University of Ibadan, Ibadan, Nigeria].

[48] Adebola MO, Amadi JE. The Efficacy of *Paecilomyces* sp. and *Penicillium digitatum* on black pod disease pathogen in the field. European Journal of Applied Science, 2012; 4[3], 101-104.

[49] Adedeji AR, Odebode AC, Agbeniyi SO. Bioassay of five

Trichoderma strains against *Phytophthora megakarya* [Cocoa pod-rot] in Nigeria. Scientific Research and Essays, 2008: 3[9], 390-394.

[50] Akrasi KO, Awuah RT. Tuber rot of yam in Ghana and evaluation of some yam rhizosphere bacteria for fungitoxicity to yam rot fungi. International Journal of AgriScience, 2012; 2[7], 571-582.

[51] Darmano TW. Capability of different isolates of Trichoderma spp. in suppressing *Phytophthora* sp. inoculum in cocoa pod tissue. Menara Perkebunan [Indonesia], 1994.

[52] Kobayashi T, Oniki M, Matsumoto K, Sitepu D, Manohara D, Tombe M, Katumoto K. Diagnostic manual for industrial crop diseases in Indonesia. Japan International Cooperation Agency and Research Institute for Spice and Medicinal Crops, 1993; Indonesia.

[53] Nyassé S, Grivet L, Risterucci AM, Blaha G, Berry D, Lanaud C, Despréaux D. Diversity of *Phytophthora megakarya* in Central and West Africa revealed by isozyme and RAPD markers. Mycological Research, 1990; 103[10], 1225-1234. doi: 10.1017/S0953756299008369

[54] Rodriguez-Kabana R, Kloepper JW, Weaver CF, Robertson DG. Control of plant parasitic nematodes with furfural-a naturally occurring fumigant. Nematropica, 1993; 63-73.

[55] Attafuah, A. Black pod disease: biocontrol. In: 1963/64 Report, Cocoa Research Institute of Ghana. 1966; Pp. 115. 1966

[56] Odamtten GT, Clerk GC. Studies on the possible use of two soil fungi, *Aspergillus niger* and *Trichoderma viride* as biological control agents of *Phytophthora palmivora* and the effect of their metabolites on cocoa seedlings. In. Proceedings International Cocoa Research Conference of Cocoa Producers' Alliance, 1985.

[57] Akrasi KO. Anti-microbial properties of yam rhizosphere micro-organisms and their potential use for controlling yam tuber rots, 2005; MSc. dissertation.

[58] Koranteng SL, Awuah RT. Biological suppression of black pod lesion development on detached cocoa pods. African Journal of Agricultural Research, 2011; 6[1], 67-72.

[59] Larbi-Koranteng S, Awuah RT, Kankam F. Biological control of black pod disease of cocoa [*Theobroma cocoa* L.] with *Bacillus amyloliquefaciens*, *Aspergillus* sp. and *Penicillium* sp. in vitro and in the field. Journal of Microbiology and Antimicrobials, 2020; 12[2], 52-63.

[60] Awuah RT. *In vivo* use of extracts from *Ocimum gratissimum* and *Cymbopogon citratus* against *Phytophthora palmivora* causing blackpod disease of cocoa. Annals of Applied Biology, 1999; 124:173-178.

[61] Amponsah, JD, Asare-Nyarko A. Glasshouse method for screening seedlings of cocoa. *Theobroma cacao* L. for resistance to black pod disease caused by *Phytophthora palmivora* (Butl.) Butl. Tropical Agriculture, 1973; 50, 143-152

[62] Abdul-Karimu A, Bosompem SA. Akumadan. *Phytophthora megakarya* resistance selection programme. Annual Report 1994; Cocoa Research Institute of Ghana, 1992/93, 58-59.

[63] Van der Vossen HAM. Selection efficiency for resistance to major diseases in cocoa. In Report on strategies of variety improvement in cocoa with emphasis on durable disease resistance. pp. 9-18. An External Review prepared for International Group for Genetic Improvement of Cocoa, UK 1997; Pp.64

[64] Nyadanu, D, Akromah, R, Adomako, B, Akrofi, AY, Dzahini-Obiatey, H, Lowor, ST, Attamah, P.

Genetic control, combining ability and heritability of resistance to stem canker in cocoa [Theobroma cocoa L.]. Euphytica, 2017; 213[12], 1-13.

[65] Nyadanu, D, Akromah, R, Adomako, B, Kwoseh, C, Lowor, ST, Dzahini-Obiatey, H, Assuah, MK. Inheritance and general combining ability studies of detached pod, leaf disc and natural field resistance to *Phytophthora palmivora* and *Phytophthora megakarya* in cocoa [*Theobroma cocoa* L.]. Euphytica, 2012; 188[2], 253-264.

[66] Ackah FK, Puije GC. van der, Moses E. First evaluation of taro [*Colocasia esculenta*] genotypes against leaf blight [*Phytophthora cololosiae*] in Ghana, HortFlora Research Spectrum, 2014; 3[4]: 390-391.

[67] Pal, KK, Gardener BM. Biological control of plant pathogens. The plant health instructor, 2006; 2, 1117-1142.

Section 2

Biocontrol of *Phytophthora* Infections

Chapter 3

Plant Beneficial Microbes Controlling Late Blight Pathogen, *Phytophthora infestans*

Brahim Oubaha, Abdellah Ezzanad and Hernando José Bolívar-Anillo

Abstract

Potato (*Solanum tuberosum*) as a food source and culinary ingredient varies is the fourth most produced noncereal crop in the world. Among multiple biotic stresses, late blight caused by *Phytophthora infestans* is the most destructive disease. Control of this pathogen is usually by the synthetic fungicides which have been fueled by the public concern about toxicity and environmental impact and development of pathogens resistance. Biological control agents (BCAs) seems the potentially alternative to these pesticides, biological disease control is now recognized and constitute an important tool in integrated pest management. BCAs strains should be able to protect the host plant from pathogens and fulfill the requirement for strong colonization. Bacteria such as Bacillus, Pseudomonas and Streptomyces and fungi such as Trichoderma and *Penicillium* were the most reported as a BCA against *P. infestans* using different direct antagonistic mode on the pathogen (via e.g. parasitism, antibiosis, or competition) or via exerting their biocontrol activity indirectly by induction in the plant of an induced systemic resistance to the pathogen. In this study, we present an overview and discussion of the use of beneficial microbes (bacteria and fungi) as novel BCAs for biocontrol of *P. infestans*.

Keywords: *Solanum tuberosum*, *Phytophthora infestans*, biological control agents, beneficial microbes

1. Introduction

Plant diseases need a good control strategies in order to maintain the quality and abundance of food around the world. Especially, human population growth has been the source of two major concerns: providing sufficient food for humanity and minimizing worldwide environmental pollution. Several approaches may be used to protect or control plant diseases. Beyond good cultural practices, harvest and post-harvest approaches in reduction of pathogen growth, growers often rely heavily on chemical fertilizers and pesticides. However, many countries have reported alarming residues of agricultural chemicals in soil, water, air, agricultural products, and even in human blood and adipose tissue [1, 2]. Additionally, research suggests that the massive use of inorganic fertilizers world-wide is associated with the accumulation of in agricultural soils [3]. Researchers and Policy makers recognize that the excessive and unsystematic application of agrichemical inputs poses a threat to the

IntechOpen

environment and humans alike. Consequently, several biologist have focused their efforts on developing alternative inputs to synthetic chemicals for controlling pests and diseases [4]. Among these alternatives those referred to as biological control by using one or more beneficial microbes to suppress the damaging activities of soil-borne pathogens.

Plant growth-promoting microbes (PGPM) are free-living microorganisms of beneficial agricultural importance. The PGPM present important beneficial effects on plant health and growth, suppress disease-causing microbes and improve nutrient. PGPM exist in the rhizosphere and this is defined as the region around the root. PGPM compensate for the reduction in plant growth caused by weed infestation [5], drought stress [6], heavy metals [7], salt stress [8, 9] and some other unfavorable environmental conditions. Beneficial microbes are also the microorganisms that produce hormones, vitamins and growth factors that improve plant growth and increase crop production. Many research reported the ability of this microorganisms to produce indole acetic acid (IAA), gibberellic acid, and cytokinins [10] and production of important metabolites such as siderophores, HCN, and antibiotics that have immense potentiality in enhancing the root surface area, altering root architecture and promoting plant growth. Among the numerous plant growth-promoting microbes (PGPM) are the most commonly applied in the biological control strategies. PGPM may affect plant performance through multiple defense mechanisms against several pathogens, operating directly by the production of specific substances that are able to promote plant growth, increase the availability and uptake of plant nutrients under biotic stress and induce the defense response of plants attacked or indirectly through the suppression of plant pathogen [11, 12].

For the biological control of late blight which is Late blight disease, caused by *Phytophthora infestans* (Mont.) de Bary, is one of the most serious threats to potato production worldwide [13], Applications of different beneficial microbes as a biocontrol bacteria, fungi, algae or their metabolites, have been tested their ability to inhibit potato late blight, and when used as part of an integrated pest management system, they have had varying degrees of success [14–16]. Bacteria with antagonistic activity toward *P. infestans* are found mainly in the genera *Pseudomonas* and *Bacillus*. Although some fungal antagonists such as *Trichoderma atroviride* and *Muscodor albus* showed effective inhibition [17–20]. The objective of this chapter is to review the ability of beneficial microbes used to control late blight of potato caused by *P. infestans*, building on recent detailed reviews and research articles on microorganisms antagonistic to late blight of potato and their management approaches.

2. The methods of isolation of *P. infestans*

P. infestans causes potato and tomato late blight, economically the most important disease of these plant species. The oomycete pathogen is frequently sampled, isolated to pure cultures and stored. Efforts were made to develop isolation and culturing techniques based on tomato and potato. There are two major steps of isolating *P. infestans*, Field collection and isolation of *P. infestans* from infected tissue [21]. Petri dish method makes easier the collection of largest number of diseased samples in the field because is based on selective medium. Leaves, stems, petioles and even slices of diseased tubers can be collected. It allows the transfer of samples from the field to the laboratory in good condition and in turn stimulating sporulation of the lesions for easy isolation. The petri dishes were prepared with 1.5-% water agar, the sample with only one lesion were chosen and placed on the plate lid with the abaxial side up, in such a way so that the agar is on the sample but never in contact. The plates must be sealed with Parafilm paper and placed in a cooler. In the laboratory, samples should be incubated at 15–18° C for 3–7 days with light and dark periods of 12 hours and grown hyphal tip of *P. infestans* transferred on a selective medium. Previous reports mentioned the use of some effectiveness selective medium for the isolation of *P. infestans*. The application of fungicides against *P. infestans* can affect the establishement of the oomycetes and their isolation. It is recommended to use tubers from fields where systemic fungicides against *P. infestans* have not been applied. Gamboa et al. [22] reported a method named sandwich method, tuber aseptically were cut in half and quickly place an infected leaflet between both halves. The both halves were attached with adhesive tape and wrap the tuber with paper towels, then place it in a paper or plastic bag for transfer to the laboratory. In the laboratory, the tuber were cuted in slices from the place of contact between the infected leaf and the tuber then put them in a wet chamber and incubate for 7 days to induce pathogen development. Incubation temperature should be 15–18° C with light and dark periods of 12 hours [22]. In the laboratory the isolation *P. infestans* from infected plant tissue can be using different infected tissues from potato or tomato plants. Sporulating lesion on potato/tomato leaves taken from field are washed in fresh water and placed in a humid chamber (inverted petri dish with water agar) with the leaf's abaxial side up and incubated at 15–18°C for 1 day or until fresh sporulation appears. Small pieces of infected tissue from the sporulating border of the lesion are cut and placed under potato/tomato slices in an empty petri dish. Dishes are incubated at 15–18°C for 1 week, until there is abundant sporulation on the upper side of the slice. To re-inoculate leaves, pick sporangia from the top of the tuber and place them in a drop of water on a potato leaf or another tuber slice. If isolating from infected tubers, slice the tuber where infection has occurred and place in a moist chamber until sporulation occurs. When clean inoculum appears on the upper side of an infected tuber slice or leaf, the sporangia are harvested in a flow chamber, by picking them up with an inoculating needle and placing the sporangia on selective medium [23, 24]. Tumwine et al. [25] reported that *P. infestans* grew successfully and well on Rye A agar without the need of antibiotics is one of the recommended medium for the isolation of *P. infestans*. The Rye A agar was described for the first time in 1968 by Caten when Rye B agar were used for the sporulation. However, V8 juice agar (V8A) which is blend of 8 vegetable juices, which supplies the trace ingredients to stimulate the growth of fungi. The acidic pH of the medium favors fungal growth and suppresses bacterial growth. V8A has been one of the most popular and commonly used medium for growth and reproduction of *Phytophthora* species [26]. In 2020, [23] were studied five different media in order to select the optimal culture conditions of *P. infestans*. Modifiations were made to use ingredients available in local markets on the following media: lime bean agar (LBA), Tree tomato or tree tomato agar (TA), carrot agar (AZ), Rye A modifid agar and 32% non-clarifid V8 agar. The findings results showed that as was described before media such as Rye A favored the ability of *P. infestans* to grow effiently.

3. Bacteria

The use of biological agents to control or suppress *Phytophthora infestans* provides an economic and environmentally friendly approach. As a biopesticides bacteria are the most common and cheaper form of microbial pesticides. The potential of a range of bacterial strains as biocontrol agents of plant pathogens has been reviewed by many scientific reports [27–29]. *Streptomyces*, *Pseudomonas* and *Bacillus* were the most tree bacteria reviewed to control *P. infestans* [30, 31]. *Actinomycetes* were isolated from in general from soil. Samples were diluted to go on serial dilution and plate on humic acid vitamin agar as described by [32] supplemented with an antifungal and antibacterial

Gram- such as nalidixic acid. The isolation plates were incubated at 35 ± 2°C for 7 days. The colonies had been transferred to International Streptomyces Project (ISP) medium No. 2 agar [33, 34] plates for purity check. This isolation method can be improved using same modifications. In the other hand, The isolation methods used to collect *bacillus* and *Pseudomonas* from soil as an endophytic or epiphytic strains were routinely grown on Luria-Bertani (LB) medium and incubated in the dark at 30°C [31, 35].

3.1 Bacillus

Bacillus and its products have been known for application as biological control agent against a range of plant pathogen. The success of *Bacillus* species as biocontrol agent could be ascribed to a wide array of peptide antibiotics produced such as iturin A, mycobacillin, subtilin and bacilysin as well as 25 different basic chemical structures with proven antifungal secondary metabolites [36, 37]. Lamsal et al. [38] found after a dual culture inhibition assay was conducted on V8-PDA in plastic petri plates (8.5 cm diameter) that seven bacterial isolates (AB05, AB11-AB15 and AB17) qualified previously as beneficial microbes of tomato plants, inhibit efficiently *P. infestans* affecting tomatoes in Korea by more than 60% in vitro. However, AB15 was the most effective, inhibiting mycelial growth of the pathogen by more than 80% in vitro. For greenhouse evaluation, targeted plants were left to dry for 2 days, and then 100 ml of bacterial spore solution (10^7 spores/ml) was added to each pot 7 days before infection so that only soil, but no above-ground parts, received any bacterial spores. The results showed that AB15 was the most effective suppressing disease by 74% compared with control plants under greenhouse conditions. According to 16S rDNA sequencing, a majority of the isolates are members of *Bacillus*, and a single isolate belongs to *Paenibacillus*. In India, for *Bacillus subtilis* strains were tested for their biocontrol activity against *P. infestans* in presence of the fungicide (Mancozeb) M45 (CURZATE®) as positive control. Before the sowing of potato seeds in blocks, all blocks were drenched with different bacterial cultures at the concentration $2x10^6$ CFU/ml, with the exception of chemical fungicide and control blocks. The potato seed tubers were treated with 0.2–0.3% of M 45 (Mancozeb) fungicide before ten days of planting. Results revealed that, bacterial treatments signifiantly reduced disease incidence of late blight compared with the control. Bacterial treatments increased the plant vegetative parameters like plant height, sprouting, number of leaves, fresh weight and dry weight of plants. In addition, treatments also showed the clear difference between commercial and non-commercial tuber yield/hectare. In a view of this results they suggest that the mode of the action were the ability of bacillus subtilis strains to produce mycotoxins which can inhibit *P. infestans* growth and the capacity of bacillus to induce the peroxidase activity [39]. Elliott et al. [40] have been reported that Companion® and Serenade® are two *Bacillus subtilis* commercial biocontrol products which reported to suppress *P. infestans*. However, resistance to this bioproducts develops and some isolates of *P. ramorum* from North American and European population have been shown to be resistant [41]. Bacillus strains could control *P. infestans* directly by inhibiting the mycelial growth, germination of the cysts or the swimming of the motile zoospore by producing many antifungal compounds which suppress the pathogen or indirectly mechanisms by inducing the inhabitation of the activity of ribosome or stimulate active oxygen burst, NO production, callose deposition, and lignification [42–44].

3.2 *Pseudomonas*

Among biocontrol agents of interest, Pseudomonas spp., are known for their production of antibiotics involved in biocontrol, such as

2,4-diacetylphluoroglucinol and phenazines [45–47], which have been widely studied in various plant-pathogen systems. Phenazine-1-carboxylic acid (PCA)-producing Pseudomonas spp., have been found effective against numerous phytopathogens, including bacteria, fungi, and oomycetes, such as the causal agent of bacterial blight of rice, *Xanthomonas oryzae* pv. oryzae [48], *Gaeumannomyces graminis* var. tritici [49] and the oomycetes *Phytophthora* spp., and Pythium spp., [50, 51]. PCA has been linked to biofilm formation, favoring attachment of PCA-producing Pseudomonas spp., to plant roots which facilitate the role of this beneficial microbes as biocontrol agents [52]. The mechanisms involved to control *P. infestans* by Pseudomonas were recently investigated, a previous study by [53] reported that the biocontrol of the pathogen could be by inhibiting sporangia and zoospore germination which suggesting the presence of many yet unknown anti-oomycete determinants. However, [54] suggests that Phenazine-1-carboxylic PCA produced by *Pseudomonas spp*., is involved in *P. infestans* growth repression and led to important transcriptomic changes by both up and down regulating gene expression in *P. infestans* over time. Different metabolic functions were altered and many effectors were found to be upregulated after the application of PCA, suggesting their implication in biocontrol. The cyclic lipopeptide surfactant massetolide A is a metabolite with versatile functions in the ecology of *Pseudomonas fluorescens* SS101 [55]. To study the effects of *P. fluorescens* SS101 and massetolide A on late blight of tomato, two leaves located on the second branch from the stem base of 5-week-old tomato plants were immersed in bacterial suspension (10^9 CFU ml^{-1}) for 1 min or in a solution of massetolide A in sterile demineralized water (pH 8). Leaves immersed in sterile demineralized water (pH 8) for 1 min served as a control. Treated tomato plants were transferred to trays covered with transparent lids. After incubation for 1 d in a growth chamber at 15°C, the lower side of each treated tomato leaf was inoculated with 3 μl droplets of a *P. infestans* zoospore suspension (3–4 × 10^3 swimming zoospores ml^{-1}) or 3 μl droplets of sterile demineralized water (pathogen-free control). *P. fluorescens* SS101 was effective in preventing infection of tomato (*Lycopersicon esculentum*) leaves by *P. infestans* and significantly reduced the expansion of existing late blight lesions. Massetolide A was an important component of the activity of *P. fluorescens* SS101, since the massA-mutant was significantly less effective in biocontrol, and purified massetolide A provided significant control of *P. infestans*, both locally and systemically via induced resistance [56]. Additionally, Biosurfactants (Rhamnolipids) produced by *fluorescent Pseudomonas* have zoospore lysis activity and biosurfactant-producing strain *Pseudomonas koreensis* 2.74 has potential to induce resistance in potato plant against late blight disease. High sensitivity of *P. infestans* zoospores to biosurfactants suggest that they can be used to dampen the spread of potato late blight once infection has been detected in the field [57, 58].

3.3 *Actinomycetes*

Actinomycetes are Gram+ bacteria that represent a high proportion of the soil microbial biomass and have the ability to produce a wide variety of antibiotics and of extracellular enzymes. Several strains of actinomycetes have been found to control plant diseases [59–61]. Recently, [62] were identifed β-rubromycin as a *P. infestans* cyst germination inhibitor by screening compounds produced by *Streptomyces* isolated from soil. For that, an acetone extract was prepared from *Streptomyces* cultures grown for 5 days in liquid medium A at 30°C by adding an equal volume of acetone followed by mixing. 20-μL aliquots were mixed with 1 × 10^3 *P. infestans* sporangia in total 70 μL (14.2% acetone solution), incubated at 10°C for 18 h, and examined using an inverted microscope. As a control, it is confirmed that 15% acetone had no effect on morphological change in *P. infestans*. The isolation of the cyst germination inhibitor enabled to identify β-Rubromycin which can inhibit *P. infestans* cyst germination and hyphal elongation from sporangia, while not affecting zoospore release, cyst formation, or appressorium formation. Chemical genetic analyses using β-rubromycin identifed a RIO kinase-like gene, PITG-04584, as a critical contributor to zoosporogenesis, cyst germination, and the formation of appressoria in *P. infestans*. The Lubimin is a vetispirane sesquiterpenoid that consists of (2R,5S,6S,8S,10R)-8-hydroxy-10-methyl-2-(prop-1-en-2-yl)spiro[4.5]decane bearing a formyl substituent at position 6. It has a role as an antifungal agent and a phytoalexin. The synthesis of this biocompounds in noninoculated potato tuber slices have been elicited after using culture filtrates of *Streptomyces* isolates which induce the resistance of potato plants against late blight caused by *P. infestans* [63]. In this sense, the reliance on actinomycetes as promising biocontrol strategies are very useful in controlling *P. infestans*. Several actnimoycetes most of which were *Streptomyces* strains have been demonstrated to be effective [64–67].

From the gastrointestinal tract of a fish dredged near the South Orkney Islands in Antarctica, [68] isolated the psychrotolerant bacterial *Vibrio splendidus* T262. Investigation of this strain led to the isolation of a rare series of 15 bis- and trisindole derivatives. Among them, six new indole alkaloids. Using the agar diffusion method, at 10 μg/paper disk, some of the isolated compounds showed activity against both gram-positive and gram-negative bacteria when trisindolal was active against the *P. infestans* and a number of other plant-pathogenic fungi.

Independently of the mode of action of biological control agents, the successful application of rhizobacteria to suppress late blight was confirmed by several research using a range of bacteria such as *Micrococcus luteus*, *Paenibacillus* spp., *Flexibacteraceae bacterium*, and *Enterobacter cloacae* [35, 40, 69, 70]. However, there is a lack of research that highlight the effectiveness of the combination assays of one or more bacteria to control *P. infestans*. Whereas, the combinations have potential for extensive colonization of the rhizosphere, more consistent expression of beneficial traits under a broad range of soil conditions, and antagonism to a larger number of pathogens than strains applied individually.

4. Fungi

The beneficial fungi have gained immense attention as biofertilizers due to their role in maintaining plant quality and quantity and their environment-friendly relationship. Nowadays, use of this microorganisms as biocontrol agent (BCA) is considered to be a rapidly developing natural phenomenon in research area. Fungal biocontrol agents (BCAs) do not cause any harm to the environment, and they generally do not develop resistance in various types of pathogens due to their complex mode of action. They have been proved to be an alternative against the undesirable use of chemical products [71–73]. Previous reports have detailed the importance of various fungi species as effectiveness biocontrol agents against *P. infestans* [74–76]. For beneficial fungi isolation, the same method was adopted for years ago based on PDA medium and it can have same small modifications. PDA with chloramphenicol 0,016% (PDAc) and Rose Bengal Agar (RBA) (dextrose 10 $g.l^{-1}$, meat peptone 10 $g.l^{-1}$, K2HPO4 1 $g.l^{-1}$, MgSO4.7H2O 0.5 $g.l^{-1}$, Rose Bengal 30 mg/l, Agar 20 g/l) media were used. Petri dishes were incubated for 4 days for bacterial isolation and 7 days for fungal isolation at 25°C in the dark [59, 77].

4.1 *Trichoderma*

In the thick of various beneficial microbes have been investigated by several scientists, Trichoderma genera is a well-known biocontrol fungi that has been used since the 1930s to help plants acquire nutrients and control the plant pathogens [78]. Several Trichoderma species have been developed commercially as biofungicides and biofertilizers.

Fungi in the genus *Trichoderma* and bacteria such as *Bacillus amyloliquefaciens* have shown in vitro potentiality to reduce the mycelial growth of *Phytophthora infetsans*, *P. quercina*, *P. capsici*, *P. cactorum* and P. plurivora attacking *Quercus robur*, *Fagus sylvatica* and *Capsicum annuum* [35, 79, 80]. The biocontrol roles of *Trichoderma* against *P. infestans* could be attributed to the Trichoderma's rhizosphere competence and competitive ability [81], via the use of many mycoparasitic strategies which are a direct mechanism for biological control that works by parasitizing, detecting, growing, and colonizing pathogen involving the detection of pathogens through chemotropism; lysis of the pathogen's cell wall, pathogen's hyphal penetration by appresorial formation; production of cell wall-degrading enzymes (CWDEs) and peptaibols and parasitizing pathogen's cell wall contents [82], antibiosis or by activating a defense response as well as increased plant growth [83]. Many studies have shown the biocontrol activity of Trichoderma against *P. infestans*. Khan et al. [84] reported for the first time the elucidation and production of viridiofungin A (VFA) from *T. harzianum* isolate T23 cultures and the antifungal potential of VFA against *P. infestans* by suppressing zoosporangia germination and exhibiting a high activity on germ-tube growth. In the assay, 0.3 ml PDB/V8 medium in 0.6 ml Eppendorf tubes containing VFA concentrations from 50 to 200 $\mu g\ ml^{-1}$ and sporangial suspensions of the pathogen were prepared. Control medium contained 2% acetone. Cultures were incubated on a shaker at 100 rpm at 25°C in the dark for 24 h. Subsequently, aliquots were taken from the cultures. Germination rates of sporangia and germ tube elongations were determined. Moreover, [85] highlighted the ability of 14 strains of *Trichoderma* to emit volatile compounds that decreased or stopped the growth of *P. infestans*. The experiments were performed in Petri plates divided into two compartments. The first compartment, containing V8 agar, was inoculated in the center with a 5 mm diameter mycelial disk of *P. infestans*. The second compartment, containing PDA, was inoculated with 5 mm mycelial disk of actively growing mycelia from one of the 14 *Trichoderma*. The plate-dividing wall prevented any physical contact between the *Trichoderma* strains and *P. infestans* but allowed the free exchange of VOCs. After inoculation, the plates were sealed with two layers of Parafilm and incubated at 21°C for 6 d, at which point the growth diameters were recorded. Volatile organic compounds (VOCs) emitted from *Trichoderma* strains inhibited the mycelial growth of *P. infestans* grown on a laboratory medium by 80% and on potato tubers by 93.1%. Using GC–MS analysis showed that the most abundant compounds were 3-methyl-1-butanol, 6-pentyl-2-pyrone, 2-methyl-1-propanol, and acetoin. Electron microscopy of the hyphae treated with *T. atroviride* VOCs revealed serious morphological and ultrastructural damages, including cell deformation, collapse, and degradation of cytoplasmic organelles.

4.2 *Penicillium*

Large number of reports mentioned that *Penicillium* spp., interact positively with plants roots. Some *Penicillium* species have shown an antagonistic activity against plant pathogens by producing antibiotics which is a primary mechanism of disease suppression by *Penicillium* also induce resistance in plants by activating defense signals [86, 87]. The adaptability to different environments and tolerance to various abiotic stresses gives theses fungi species an advanced ranking to suppress many plant pathogens [87]. Previous reports have demonstrated that *Penicillium* species show efficacy as biocontrol agent against *P. infestans*. Based on the study conducted by [77] reported that *Penicillium chrysogenum* induce resistance against *P. infestans* in tomato plants. Dry *Peni. chrysogenum* mycelium extract was prepared using a detailed protocol described by [77] the extract was diluted with distilled water to a total carbohydrate content of 1.5 g l^{-1}. The tomato plants were treated two times with about 25 mL extract per plant as foliar spray. Leaf discs (diam. 18 mm) of plants treated were laid onto moist filter paper. Leaf discs were inoculated with 10 L droplets of zoospore suspension. The inoculated leaf material was kept at 23°C in the dark with a relative humidity at 100%. Three days later biochemical assays for the peroxidase activity and isoenzyme analysis were conducted. The application of the water extract of killed *Peni. chrysogenum* has shown no direct antifungal activity against the pathogen, however the protective effect of the extract was shown under controlled conditions after application on the whole plants and on leaf disk. The findings suggest that control resulted from the induction of defense mechanisms in the tomato plants. According to this many reports have been shown that the ability of *Penicillium* species to induce systemic expressions of defense-related genes [peroxidases (POX) and phenylalanine ammonia lyase (PAL) and PR-1 genes] is the key used by *Penicillium* to induce plant defense systems as well protects plants from pathogens [85, 86]. Otherwise, antagonistic activity of endophytic fungi associated with *Artemisia nilagirica* was studied against the pathogen *P. infestans* by the presence or absence of inhibition zone observed in dual cultures by using dual culture methods. The study has shown that among the endophytic fungal tested *Penicillium atrovenetum* and *Trichoderma viride* showed direct inhibition activity of pathogen mycelia growth [87]. Additionally, [88] reported that *P. striatisporum* Pst10 isolated from the rhizosphere of chili peppers showed very high antagonistic effects on mycelium growth and sporangia/spore formation or germination of *Phytophthora* spp., The analysis of the Pst10 organic solvent extract by thin-layer chromatography (TLC) and the antagonistic activity tests highlight the existence of three different antifungal compounds produced by *P. striatisporum* Pst10. To study the Pst10 antifungal spectrum of Pst10 the dual culture assays were used. In the other hand, To determine the effect of Pst10 sterilized liquid culture filtrates (SLCF) on sporangium and spore germination, 100 μl of sporangium or spore suspensions of *P. capsici* were spread on 20 ml PDA agar containing 1 ml Pst10 SLCF. PDA plates were incubated at 28 C for 24, 72, and 120 h. After each incubation time, 100 sporangia or spores were counted and germination rate was calculated under a light microscope.

Using fungi as biological agents to control or suppress the growth of *P. infestans* is not just limited to *Trichoderma* and *Penicillium* even they were the most fungi reported. In 2020 [75] Isolated *Aspergillus flavipes* from agricultural soils as a strong inhibitor for growth of various species of *Phytophthora*. As well as, the crude extracellular extract of broth cultures of *A. flavipes* displayed a significant growth inhibition of various *Phytophthora spp.*, The putative compounds from *A. flavipes* were chemically verified as 3-hydroxy-2′,4,4′,6′-tetramethoxychalocone, 7,3,4,5′-tetramethoxyflavanone, isovitexin and amodiaquine. The non-activity of this compounds on several pathogens while their noticeable drastic effect on *Phytophtora* zoospores germination, mycelial anastomosis, sporangial formation and causing enlarged hyphal tips, dwarfness to the hyphal length. This results suggest that *A. flavipes* compounds are considered potentially as antiphytophthoral. Moreover, [89] described an antifungal metabolite, oosporein, which was isolated from the liquid culture of *Verticillium psalliotae* that produced the antagonistic effects on *P. infestans*. Oosporein exhibited a significant growth-inhibitory effect on *P. infestans* in comparison with other phytopathogenic fungi. De Vries et al. [90] found that Out of an analysis of 12 fungal endophytes, *Phoma eupatorii* isolate 8082 and *Monosporascus* spp., inhibited the growth of *P. infestans* in co-culture using the agar diffusion assays, co-inoculation in planta and anthocyanin, presumably through the secretion of secondary metabolites, particularly since their culture extracts were also active. Furthermore, the study reported that the two of the endophytes exhibited global inhibition of nine European *P. infestans* isolates. These examples indicate that many fungi species as a beneficial microbes are also characterized with high potential to control *P. infestans* directly by antagonistic activity which inhibit the mycelia growth and the zoospore/zoosporangia germination via the production of a range of biocompounds and by the induction of defense mechanisms. Nonetheless, the use of beneficial fungi as a potential candidate to be more studied and tested as a novel biocontrol agent in the field providing an alternative to resistance gene breeding and application of agrochemicals.

5. Mode of action

As mentioned early, previous investigations highlight the importance of fungi and bacteria as biological control against *P. infestans*. Thus, gaining insight into mechanisms is of high importance for disease control. It is reported that microorganisms engage several antagonistic mechanisms against plant pathogens, including antibiosis, mycoparasitism, competition for nutrients and space, promotion of plant growth, induced plant defense mechanisms, and modification of environmental conditions. Among those mechanisms, the antibiosis refers to interaction lethal between microorganisms through secondary metabolites, which is of high importance to identify target cell, protein or enzyme, in concrete, implicated in the mechanism. Moreover, identification of chemical substance responsible on inhibiting of plant pathogens is a task challenge, due to volatility of compounds and their synergetic effects. Until now, fewer compounds from microorganisms were shown to effectively affect *P. infetans*. These include Phenazine-carboxylic acid [91], Oosporein [92], β-Rubromycin [93], Iturin A [94], Fenngycin A' [95, 96], Thiobutacin [97], Bikaverin [98], Fusaric acid [98], 2,5-diketopiperazine [99] and Xenocumacine 1 [100], listed in **Figures 1** and **2**. Moreover, detailed mechanism of interaction against *P. infestans* was developed only with β-Rubromycin, Iturin A and phenazine-1-carboxylic acid. β-Rubromycin belongs to the quinone antibiotics that have the ability to inhibit retroviral reverse transcriptase but also act as inhibitors of DNA polymerases [101]. [94] evaluated the activity of β-Rubromycin produced by *Streptomyces* isolated from soil against *P. infestans*, showing that this compound was capable of inhibiting the infection caused by sporangia and zoospores in tomato plants. The mechanism of action seems to be related to the up regulation of the RIO kinase-like gene that are involved in morphological development, altering processes as important in *P. infestans* as cyst germination and hyphal elongation. [95] evaluated the biocontrol capacity of *Bacillus subtilis* WL-2 against *P. infestans*, establishing that Iturin A was the metabolite involved in the inhibition capacity against this phytopathogen, causing cell membrane disruption and an irregular internal cell structure. Iturin A is a lipopeptide that exerts its antimicrobial action through the alteration of the cell membrane via the production of pores that generate osmotic perturbation [102]. In addition to its activity in the membrane it was observed that iturin A was capable of generating mitochondrial damage in *P. infestans*, causing oxidative stress and alterations in the respiratory chain which alter ATP synthesis. [54] reported the effect of phenazine-1-carboxylic acid (PCA) produced by a strain of *Pseudomonas fluorescens* on the transcriptome of *P. infestans*, establishing that this

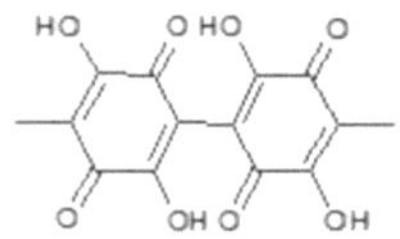

Oosporein
(from *Verticillium psalliotae*)

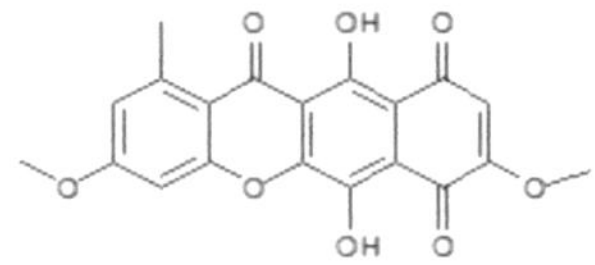

Bikaverin
(from *Fusarium oxysporum*)

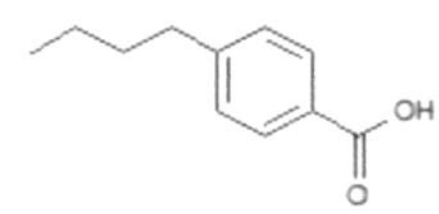

fusaric acid
(from *Fusarium oxysporum*)

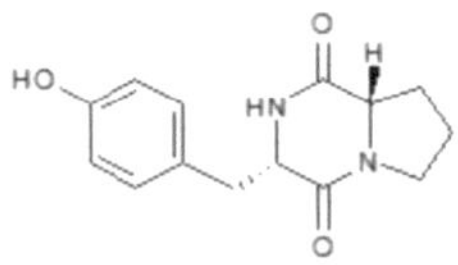

Cyclo (L-Pro-L-Tyr)
2,5-diketopiperazine
(from *Lysobacter capsici AZ78*)

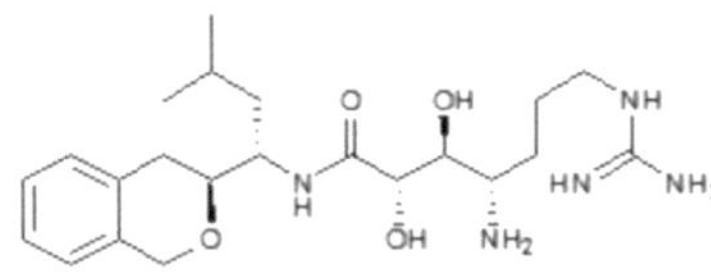

Xenocoumacin 1
(from *Xenorhabdus nematophilus var. pekimgensis*)

Figure 1.
*Anti-*Phytophthora infestans *compounds produced by fungi microorganisms.*

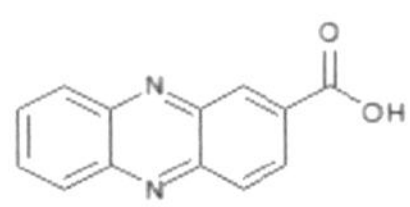

Phenazine-1-carboxylic acid
(from Pseudomonas fluorescens strain LBUM636)

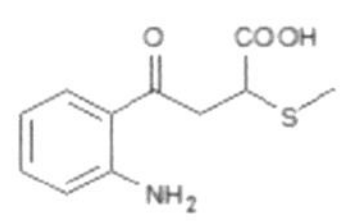

Thiobutacin
(from *Lechevalieria aerocolonigenes*)

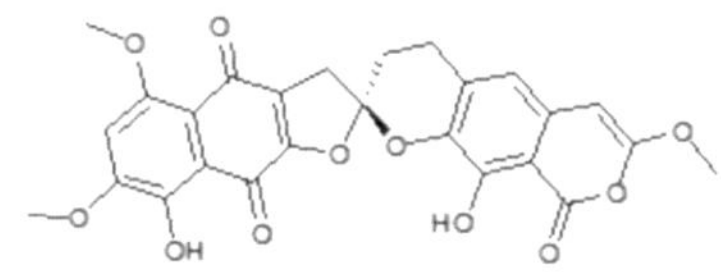

b-Rubromycin
(from *Streptomyces sp.* no. 750)

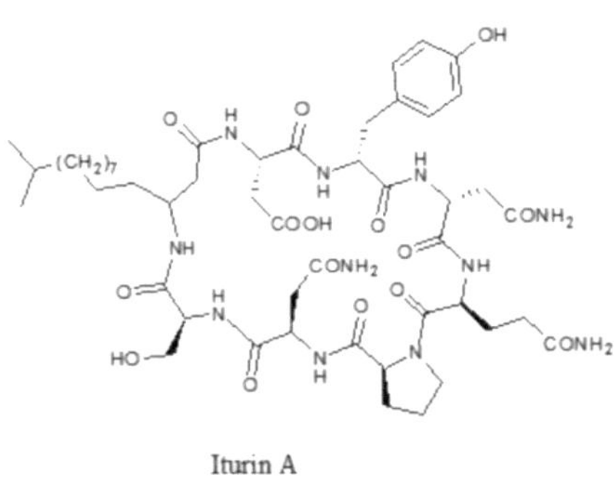

Iturin A
(from Bacillus subtilia WL-2)

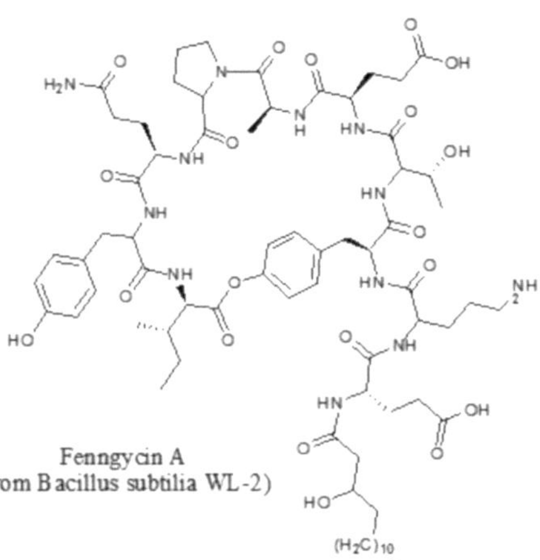

Fenngycin A
(from Bacillus subtilia WL-2)

Figure 2.
*Anti-*Phytophthora infestans *compounds produced by bacteria microorganisms.*

compound alters the expression of genes involved in functions like phosphorylation mechanisms, transmembrane transport and oxydo-reduction activities.

Another method of disease control, so-called Mycroparasitism, is able to antagonize plant pathogens and promote plant growth by treatment with other microorganisms. Mycoparasitism is a direct mechanism in which microogranism colonizes the pathogen through detection, parasitization and growth actions [103, 104]. This protection strategy has been recognized as an important mechanism of biological control. Mycoparasitics such as the oomycete *Pythium oligandrum* [105], *Pythium periplocum* [106] and different species from *Trichoderma* including *T. asperellum*, *T. atroviride*, *T. virens*, and *T. harzianum* are successfully used against *P. infetans*. These mycoparasitic grow faster than their pathogenic plant counterparts, which means that they can occupy rhizosphere space and nutrition, thus promoting both plant growth [107] resistance in host plants [108, 109]. The mechanism of Trichoderma spp., for example, appear to be very complex involving the detection of plant pathogen through chemotropism; lysis of the pathogen's cell wall (the key to mycoparasitism) [110]; pathogen's hyphal penetration by appresorial formation; production of cell wall-degrading enzymes (CWDEs) and peptaibols, mediated by heterotrimeric G-proteins and mitogen-activated protein (MAP) kinases [111]; and parasitizing pathogen's cell wall contents [112]. Degradation of pathogen's cell wall during mycoparasitism is mediated by a set of hydrolytic enzymes including β-(1,6)-glucanases, chitinases, and proteases. Several members from each of these classes have been shown to be involved in mycoparasitism and/or to be induced under mycoparasitism-related growth conditions [113]. Although these microorganisms demonstrate their potential as mycoparasitic biological control agents, fewer mechanistic studies have been investigate the molecular or genetic determinants of their mycoparasite lifestyle.

However, rather than directly expanding into infected plant, microorganism might compete with the pathogen producing secondary metabolites able to partially or totally inhibit the pathogenic fungi. This classical mechanism occurs when special and nutritious resources are limited. Consequently, the antagonistic microorganisms feed on the available resources for growth, causing therefore a reduction in the growth of the pathogens. A published example of metabolite-pathogen protection is that produced by *Phoma eupatorii* 8082. This endophyte has a remarkable potential to produce the anthocyanin product [114]. The latter could be produced as a result of a stress response positively regulated by jasmonic acid [115–118]. Hence, it is possible that tissue colonization with *Pho. Eupatorii* induce jasmonic acid dependent defense responses, which may play a role in the inhibition of the *P. infestans* infection. Indeed, [119] reported that jasmonic acid induced reduction of infection in the leaves of tomato and potato plants and [120] testified the mandatory existence of jasmonic acid to activate the defensive responses elicited by a peptide secreted by *P. infestans*. Some microorganisms including *Trichoderma* spp., produce inorganic compounds able to alter soil pH and therefore able to modify micronutrients (phosphate, iron and Manganese) [121]. In these extreme conditions *Trichoderma* spp., were able to produce various kinds of Siderophore products [122], including: caprogens, ferrichromes and fusarinines [123], thanks to the change in non-ribosomal peptide synthetase products and diverse non-ribosomal peptide synthetase-encoding genes [124]. Siderophores play a dual role, an antagonistic agent by inhibiting or even suppressing the growth of pathogens by divesting source of iron, as well as an agonist agent that helps to solubilize iron that was not available to the plant. These abilities explain the competition mechanism on the nutrient resources.

Alternative mechanism of disease control against attack of pathogens is based on the induction of systemic and local resistances [125]. Such resistances result from complex interactions between plants and antagonist elicitors, provoking physiological and biochemical alteration of cells. Indeed, two major kinds of systemic resistances have been studied; systemic acquired resistance (SAR) [126] and induced systemic resistance (ISR) [127]. Both systemic resistances are based on distinct phytohormonal signals. Various compounds have been proposed as potential signals for systemic resistances activation. The non-protein amino acid, β-Aminobutyric acid

(BABA), is known to induce resistance against various pathogens on a wide range of plants. Indeed, DL- β-Aminobutyric acid-induced resistance of potato against late blight pathogen *P. infestans* trough the signaling compound salicylic acid [128]. BABA also provided significant control against *P. infestans* on tomato [129]. The systemic defense is induced in a salicylic acid dependent manner; furthermore various inorganic chemicals including indole acetic acid, di-potassium hydrogen orthophosphate, hydrogen peroxide, calcium chloride, ferric chloride and metalaxyl were able to induce resistance against the disease caused by *P. infestans*. Treatment with those agents promotes the synthesis of defense enzymes like peroxidase, polyphenol oxidase (POX) and phenylalanine ammonia lyase (PAL) [130]. In addition, Curdlan b-1,3-Glucooligosaccharides has shown to enhance plant resistance against the pathogen *P. infestans* in foliar tissues of potato (*Solanum tuberosum* L. cv. McCain G1) by accumulation of H2O2 and salicylic acid and the activities of phenylalanine amino-lyase, b-1,3-glucanase and chitinase [131].

6. Conclusion

The application of beneficial bacteria and fungi as biocontrol agents is an interesting building block of sustainable and environmentally sound management strategies of *Phytophtora infestans*. A holistic approach should be considered to reach satisfactory levels of *P. infestans* control by a beneficial microbes. Based on the number of currently known isolates with biocontrol activity against *P. infestans*, the predominant genera are *Pseudomonas*, *Bacillus*, *Streptomyces*, *Trichoderma* and *Penicillium*. The ability to affect survival structures, sharing the same ecological niche as *Phytophtora*, inducing resistance responses in the plant and promoting plant growth are desirable characteristics of a competent BCA against *P. infestans*. However, among several criteria the potential bottlenecks such as large-scale production, formulation, preservation conditions, shelf life, application methods, and combination potentiality of one or more microbes should be tackled early in the selection process.

Conflict of interest

The authors declare no conflict of interest.

Appendices and nomenclature

BABA	Acide bêta-aminobutyrique.
BCA	Biological control agent.
CWDE	Cell wall degrading enzymes.
HCN	Hydrogen cyanide.
ISR	Induced systemic resistance.
IAA	Indole acetic acid.
MAP	Mitogen-activated protein.
PAL	Phenylalanine ammonia lyase.
PCA	Phenazine-1-carboxylic acid.
PGPM	Plant growth promoting microbes.
POX	Peroxidase.
PR-protein	Pathogenesis related protein.
SAR	Systemic acquired resistance.

TLC	Thin layer chromatography.
VFA	Virdiofungin A.
VOC	Volatile organic compounds.

Author details

Brahim Oubaha[1]*, Abdellah Ezzanad[2] and Hernando José Bolívar-Anillo[3]

1 Laboratory of Microbial Biotechnologies Agrosciences and Environment (BioMAgE), Department of Biology, Faculty of Sciences Semlalia, University of Cadi Ayyad, Marrakech, Morocco

2 Department of Organic Chemistry, Faculty of Sciences, University of Cadiz, Cadiz, Spain

3 Laboratorio de Investigación en Microbiología, Facultad de Ciencias Básicas y Biomédicas, Universidad Simón Bolívar, Barranquilla, Colombia

*Address all correspondence to: oubaha123@gmail.com; brahim.oubaha@edu.uca.ac.ma

References

[1] Udeigwc TK, Teboh JM, Eze PN. Implications of leading crop production practices on environmental quality and human health. J Environ Manage. 2015;151:267-279.

[2] Alvarez A, Saez JM, Costa JSD. Actinobacteria: current research and perspectives for bioremediation of pesticides and heavy metals. Chemosphere 2017;166:41-62.

[3] Ridolfi AS, Alvarez GB, Rodríguez Giraul ME. Organochlorinated contaminants in general population of Argentina and other Latin American Countries. In Alvarez A, Polti M (eds). Bioremediation in Latin America. Cur Res Persp. 2014;134:17-40.

[4] Carey AE, Kutz FW. Trends in ambient concentrations of agrochemicals in humans and the environment of the United States. Environ Monit Assess 1985;5:155-163.

[5] Babalola OO, Sanni AI, Odhiambo GD, Torto B. Plant growth-promoting rhizobacteria do not pose any deleterious effect on cowpea and detectable amounts of ethylene are produced. World J Microbiol Biotechnol. 2007;23(6):747-752.

[6] Zahir ZA, Munir A, Asghar HN, Shaharoona B, Arshad M. Effectiveness of rhizobacteria containing ACC deaminase for growth promotion of peas (*Pisum sativum*) under drought conditions. J Microbiol Biotechnol. 2008;18(5):958-963.

[7] Kumar KV, Srivastava S, Singh N, Behl HM. Role of metal resistant plant growth promoting bacteria in ameliorating fly ash to the growth of *Brassica juncea*. J Hazard Mater. 2009;170(1):51-57

[8] Egamberdieva D. Plant growth promoting properties of rhizobacteria isolated from wheat and pea grown in loamy sand soil. Turk J Biol. 2008;32(1): 9-15

[9] Kaymak HC, Guvenc I, Yarali F, Donmez MF. The effects of bio-priming with PGPR on germination of radish (*Raphanus sativus* L.) seeds under saline conditions. Turk J Agric For. 2009;33(2): 173-179.

[10] Kloepper JW, Gutierrez-Estrada A, Mclnroy JA. Photoperiod regulates elicitation of growth promotion but not induced resistance by plant growth-promoting rhizobacteria. Can J Microbiol. 2007;53:159-167.

[11] Ribeiro CM, Cardoso, EJBN. Isolation, selection and characterization of root-associated growth promoting bacteria in Brazil Pine (*Araucaria angustifolia*). Microbiol Res. 2012;167:69-78. Doi: 10.1016/ j.micres.2011.03.003

[12] Oubaha B, Nafis A, Ezznad A, Stump M, Félix M, Barakate M. Plant growth promotion and biological control of *Aphanomyces euteiches*, causal agent of rot root in legumes, by Moroccan actinobacteria J Biocont Sci Tech. 2020;30(11)127-143. DOI: 10.1080/09583157.2020.1846160

[13] Taylor RJ, Pasche JS, Gallup CA, Shew HD, Gudmestad NC. A foliar blight and tuber rot of potato caused by *Phytophthora nicotianae*: New occurrences and characterization of isolates. Plant Dis. 2008;92:492-503.

[14] Yan ZN, Reddy MS, Ryu CM, McInroy JA, Wilson M, Kloepper JW. Induced systemic protection against tomato late blight elicited by plant growth-promoting rhizobacteria. Phytopathol. 2002;92:1329-1333.

[15] Xel C, Zannini E, Coffey A, Guo J, Waters DM, Arendt EK. Ecofriendly

control of potato late blight causative agent and the potential role of lactic acid bacteria: a review. Appl. Microbiol. Biotechnol. 2012;96:37-48.

[16] Gachango, E.; Kirk, W.; Schafer, R.; Wharton, P. Evaluation and comparison of biocontrol and conventional fungicides for control of postharvest potato tuber diseases. Biol Control. 2012;63:115-120.

[17] Daayf F, Adam L, Fernando WGD. Comparative screening of bacteria for biological control of potato late blight (strain US-8), using in-vitro, detached-leaves, and whole-plant testing systems. Can. J. Plant Pathol. 2003;25:276-284.

[18] Kloepper JW, Ryu CM, Zhang SA. Induced systemic resistance and promotion of plant growth by *Bacillus* spp. Phytopathol. 2004;94:1259-1266.

[19] Sanchez V, Bustamante E, Shattock R. Selection of antagonists for biological control of *Phytophthora infestans* in tomato. Manejo Integrado de Plagas. 1998;48:25-34.

[20] Stoppacher N, Kluger B, Zeilinger S, Krska R, Schuhmacher R. Identification and profiling of volatile metabolites of the biocontrol fungus *Trichoderma atroviride* by HS-SPME-GCMS. J Microbiol Methods. 2010;81:187-193.

[21] Sato N, Kato M, Mosa A, Kobayashi K, Ogoshi A. A newspaper bag method for sample collection of blighted potato leaflets for isolation of *Phytophthora infestans*. Ann Phytopath Soc Japan. 1991;57:573-576.

[22] Gamboa S, Perez W, Andrade-Piedra JL, Forbes G. 2019. Laboratory manual for *Phytophthora infestans* work at CIP. Lima, Peru: International Potato Center: 78pp.

[23] Gómez-González S, Castañeda-Sánchez D, Morales-Osorio *J. media* preferences, micro-morphometric analysis, and cardinal growth temperature determination for *Phytophthora infestans* sensu lato isolated from different hosts in Colombia. Braz J Bio. 2019;(80)1:167-179. DOI:10.1590/1519-6984.206460

[24] Sobkowiak S, Śliwka J. *Phytophthora infestans*: isolation of pure cultures, storage and inoculum preparation. Plant Breeding Seed Sci. 2017;(76)1:9-15. DOI: 10.1515/plass-2017-0015

[25] Tumwine J, Frinking H, Jeger J. Isolation techniques and cultural media for *Phytophthora infestans* from tomatoes. Mycologist. 2000;(14)3: 137-139.

[26] Erwin D.C, Ribeiro K. *Phytophthora* diseases worldwide. St. Paul: APS Press, The Ame Phytopathol Soc. 1996.

[27] Eljounaidi K, Lee SK, Bae H. Bacterial endophytes as potential biocontrol agents of vascular wilt diseases - Review and future prospects. Biol Control. 2016;103:62-68. DOI: 10.1016/j.biocontrol.2016.07.013

[28] Oubaha B, Nafis A. *Aphanomyces euteiches*, the oomycete responsible for peas' root rot-current views on the methods for inhibiting its pathogenicity. J Appl Sci Envir Stud. 2020;3(3):147-161. http://revues.imist.ma/index.php?journal=jases

[29] Oubaha B, Ezzanad A, Bolivar-Anillo HJ. Aperçu sur *Aphanomyces euteiches*; agent causal de fonte de semis et la pourriture racinaire chez le petit pois, Biologie et méthodes de lutte. Editions Universitaires Européennes (EUE); 2021. 24 p. ISBN-13: 978-620-3-41344-1, ISBN-10: 6203413445.

[30] Chumpitaz A, Caro J, Cruz W, León J. Rhizospheric actinomycetes from organic crops of native potato (*Solanum tuberosum*): isolation, phenotypic characterization, molecular identification, and impact on biocontrol

of *Phytophthora infestans* (Mont.) de Bary. Scientia Agropecuaria. 2020;11(2):223-231.

[31] Caulier S, Gillis A, Colau G, Licciardi F, Liépin M, Desoignies N, Modrie P, Legrève A, Mahillon J and Bragard C. Versatile Antagonistic Activities of Soil-Borne Bacillus spp. and *Pseudomonas* spp., against *Phytophthora infestans* and Other Potato Pathogens. Front Microbiol. 2018;9:143-162. DOI: 10.3389/fmicb.2018.00143

[32] Hayakawa M, Nonomura H. Humic acid-vitamin agar, a new medium for the selective isolation of soil *actinomycetes*. J Ferment Technol. 1987;65:501-509.

[33] Shirling EB, Gottlieb D. Methods for characterization of *Streptomyces* species. Int J Syst Evol Microbiol. 1966;16:313-340.

[34] Khucharoenphaisan K, Puangpetch U, Puttaraksa K, Sinma K. Grouping of actinomycetes isolated from termites using biochemical character. J Biol Sci. 2011;11:314-319.

[35] Bruisson S, Zufferey M, L'Haridon F, Trutmann E, Anand A, Dutartre A, De Vrieze M, Weisskopf L. Endophytes and epiphytes from the grapevine leaf microbiome as potential biocontrol agents against phytopathogens. Front Microbiol. 2019;10:2726-2741. DOI: 10.3389/fmicb.2019.02726

[36] Yoshida S, Hiradate S, Tsukamot T, Hatakeda K, Shirata A. Antimicrobial activity of culture fitrate of *Bacillus amyloliquefaciens* RC-2 isolated from Mulberry leaves. Phytopathol. 2000;91:181-187.

[37] Mutaz MA, Hasnain S. Simple and rapid isolation of a novel antibiotic from *Bacillus subtilis* Mz-7. J Liq Chromatogr Relat Technol. 2006;29:639-647.

[38] Lamsal K, Kim SW, Kim YS, Lee YS. Biocontrol of Late Blight and Plant Growth Promotion in Tomato Using Rhizobacterial Isolates. J. Microbiol. Biotechnol. 2013;23:897-904. DOI: 10.4014/jmb.1209.09069

[39] Kumbar B, Mahmood R, Nagesha SN, Nagaraja MS, Prashant DG, Zablon O. Field application of *Bacillus subtilis* isolates for controlling late blight disease of potato caused by *Phytophthora infestans*. Biocatal Agricult Biotechnol. 2019;22:13-56. DOI:10.1016/j.bcab. 2019.101366

[40] Elliott M, Shamoun SF, Sumampong G, James D, Masri S, Varga A. Evaluation of several commercial biocontrol products on European and North American populations of *Phytophthora ramorum*, Biocontrol Sci Technol, 2009;19 (10):1007-1021. DOI: 10.1080/ 09583150903243870

[41] Wagner S, Kaminsky K, Werres S. Evaluation of Fungicides for Control of *Phytophthora ramorum*. In: Proceedings of the Sudden Oak Death Third Science Symposium, CA: Pacific Southwest Research Station, Forest Service, US Department of Agriculture; 12-15 June 2008; New York: PSOD; 2008. P.491-495.

[42] Chowdhury SP, Hartmann A, Gao X. Borriss R. Biocontrol mechanism by root-associated *Bacillus amyloliquefaciens* FZB42–a review. Front Microbiol. 2015;6:780-796.

[43] Lu X, Zhou DM, Chen X, Zhang JF, Huang HW, Wei LH. Isolation and characterization of *Bacillus altitudinis* JSCX-1 as a new potential biocontrol agent against *Phytophthora sojae* in soybean. Plant Soil. 2017;416:53-66.

[44] Liu D, Li K, Hu J, Wang W, Liu X, Gao Z. Biocontrol and Action Mechanism of *Bacillus amyloliquefaciens* and *Bacillus subtilis* in Soybean *Phytophthora* Blight. Int J Mol Sci.

2019;20(12):2908-2921. DOI: 10.3390/ijms20122908

[45] Loper JE, Hassan KA, Mavrodi DV, Davis EW, Lim CK, Shaffer BT. Comparative genomics of plant-associated *Pseudomonas* spp.: insights into diversity and inheritance of traits involved in multitrophic interactions. PLoS Genet. 2012;8(7):84-111. DOI: 10.1371/journal.pgen.1002784

[46] Mavrodi DV, Mavrodi OV, Parejko JA, Bonsall RF, Kwak YS, Paulitz TC. Accumulation of the antibiotic phenazine-1-carboxylic acid in the rhizosphere of dryland cereals. Appl Environ Microbiol. 2012;78: 804-812.

[47] Morohoshi T, Wang WZ, Suto T, Saito Y, Ito S, Someya N, Ikeda T. Phenazine antibiotic production and antifungal activity are regulated by multiple quorum-sensing systems in *Pseudomonas chlororaphis subsp. Aurantiaca* StFRB508. J Biosci Bioeng. 2013;116:580-594.

[48] Xu S, Pan X, Luo J, Wu J, Zhou Z, Liang X, He Y, Zhou M. Effects of phenazine-1-carboxylic acid on the biology of the plant-pathogenic bacterium *Xanthomonas oryzae pv. oryzae*. Pestic Biochem Phys. 2015;117:39-46.

[49] Thomashow LS, Weller DM. Role of a phenazine antibiotic from *Pseudomonas fluorescens* in biological control of *Gaeumannomyces graminis var. tritici*. J Bacteriol. 1988;170:3499-3508.

[50] Puopolo G, Masi M, Raio A, Andolfi A, Zoina A, Cimmino A, Evidente A. Insights on the susceptibility of plant pathogenic fungi to phenazine-1-carboxylic acid and its chemical derivatives. Nat Prod Res. 2013;27:956-966.

[51] Gurusiddaiah S, Weller D, Sarkar A, Cook R. Characterization of an antibiotic produced by a strain of *Pseudomonas fluorescens* inhibitory to *Gaeumannomyces graminis var. tritici* and *Pythium* spp. Antimicrob Agents Ch. 1986;29:488-495.

[52] Mavrodi DV, Blankenfeldt W, Thomashow LS. Phenazine coumpounds in fluorescent *Pseudomonas* spp. biosynthesis and regulation. Annu Rev Phytopathol. 2006;13:417-445.

[53] De Vrieze M, Varadarajan AR, Schneeberger K, Bailly A, Rohr RP, Ahrens CH, Weisskopf L. Linking comparative genomics of nine potato-associated *Pseudomonas* isolates with their differing biocontrol potential against late blight. Front. Microbiol. 2020;11(857):2138-2158. DOI: 10.3389/fmicb.2020.00857

[54] Roquigny R, Novinscak A, Arseneault T, Joly DL, Filion M. Transcriptome alteration in *Phytophthora infestans* in response to phenazine-1-carboxylic acid production by *Pseudomonas fluorescens* strain LBUM223. BMC Genomics, 2018;19(1):474-489. DOI: 10.1186/s12864-018-4852-1

[55] De Souza JT, de Boer M, De Waard P, Van Beek TA, Raaijmakers JM. Biochemical, genetic, and zoosporicidal properties of cyclic lipopeptide surfactants produced by *Pseudomonas fluorescens*. App Envir Microbiol. 2003;69:7161-7172.

[56] Tran H, Ficke A, Asiimwe T, Höfte M, Raaijmakers JM. Role of the cyclic lipopeptide massetolide A in biological control of *Phytophthora infestans* and in colonization of tomato plants by *Pseudomonas fluorescens*. New Phytologist. 2007;175(4):731-742. DOI: 10.1111/j.1469 8137.2007.02138.x

[57] Bengtsson TE. Liljeroth E, Andreasson A, Holefors M, Hultberg N. Biosurfactants have the potential to induce defence against *Phytophthora*

infestans in potato. Potato Res. 2015;58:83-90.

[58] Tomar S, Lal M, Khan MA, Singh BP Sharma S. Characterization of glycolipid biosurfactant from *Pseudomonas aeruginosa* PA1 and its efficacy against *Phytophtora infestans*. J. Environ. Biol. 2019;40:725-730.

[59] Erginbas G, Yamac M, Amoroso MJ, Cuozzo SA. Selection of antagonistic *actinomycete* isolates as biocontrol agents against root-rot fungi. Fresenius Enviro Bull. 2010;19:11-19.

[60] Oubaha B, Nafis A, Baz M, Mauch F, Barakate M. The potential of antagonistic Moroccan *Streptomyces* isolates for the biological control of damping-off disease of pea (*Pisum sativum* L.) caused by *Aphanomyces euteiches*. J Phytopathol. 2018;167:82-90. DOI: 10.1111/jph.12775

[61] Shariffah-Muzaimah SA. Idris AS, Madihah AZ, Dzolkhifli O, Kamaruzzaman S, Maizatul-Suriza M. Characterization of Streptomyces spp. Isolated from the rhizosphere of oil palm and evaluation of their ability to suppress basal stem rot disease in oil palm seedlings when applied as powder formulations in a glasshouse trial. W J Microbiol Biotechnol. 2018;34(15):476-487. DOI: 10.1007/s11274-017-2396-1

[62] Tani S, Nishio N, Kai K, Hagiwara D, Ogata Y, Tojo M, Sumitani Y, Judelson S, Kawaguchi T. Chemical genetic approach using β-rubromycin reveals that a RIO kinase-like protein is involved in morphological development in *Phytophthora infestans*. Sci Rep. 2020;10(26):223-236. DOI: 10.1038/s41598-020-79326-7

[63] Keqiang C, Forrer H. Current status and prosperity on biological control of potato late blight (*Phytophthora infestans*). J Agri Univ Hebei. 2001;26(12):436-451.

[64] Ardila Y, Castellanos D, Sicard T. Efecto Antagónico in vitro de Actinomicetos Aislados de Purines de Chipaca (*Bidens pilosa* L.) Frente a *Phytophthora infestans* (Mont) de Bary. Rev Fac Nal Agr Medellín. 2011;64(2):6111-6119.

[65] Khucharoenphaisan K., Rodbangpong K., Saengpaen P., Sinma K. (2016) Exploration on soil *actinomycetes* against *Phytophthora* sp. causing root rot of cassava and plant growth promoting activities. J Plant Sci. 2016;11:38-44. DOI: 10.3923/jps.2016.38.44

[66] Cray JA, Connor MC, Stevenson A, Houghton JDR, Rangel DEN, Cooke LR, Hallsworth JE. Biocontrol agents promote growth of potato pathogens, depending on environmental conditions. Microbial Biotechnol. 2016;9(3):330-354. DOI:10.1111/1751-7915.12349

[67] Marfetán JA, Greslebin AG, Taccari LE, Vélez M. Rhizospheric microorganisms as potential biocontrol agents against *Phytophthora austrocedri*. Eur J Plant Pathol. 2020;158 :721-732. DOI: 10.1007/s10658-020-02113-7

[68] Nair V, Schuhmann I, Anke H, Kelter G, Fiebig HH, Helmke E, Laatsch H. Marine bacteria, XLVII – Psychrotolerant bacteria from extreme antarctic habitats as producers of rare bis- and trisindole alkaloids. Planta Medica, 2016;82(10):910-918. DOI:10.1055/s-0042-108204

[69] Patricia J, Slininger M, David A, Schisler L, Ericsson LD, Brandt T, Jo Frazier M, Woodell LK, Olsen NL, Kleinkopf GE. Biological control of post-harvest late blight of potatoes, Biocontrol Sci Technol. 2007;17:(6)647-663. DOI: 10.1080/09583150701408881

[70] Bruisson S, Zufferey M, L'Haridon F, Trutmann E, Anand A, Dutartre A, De Vrieze M, Weisskopf L.

Endophytes and epiphytes from the grapevine leaf microbiome as potential biocontrol agents against phytopathogens. Front Microbiol. 2019;10:2726-2739. DOI: 10.3389/fmicb.2019.02726

[71] El-Sharkawy HHA, Rashad YM, Ibrahim SA. Biocontrol of stem rust disease of wheat using arbuscular mycorrhizal fungi and *Trichoderma* spp. Physiol Mol Plant Pathol. 2018;103:84-91.

[72] Zhou L, Tang K, Guo S. The plant growth-promoting fungus (PGPF) *Alternaria* sp. A13 markedly enhances *Salvia miltiorrhiza* root growth and active ingredient accumulation under greenhouse and field conditions. Int J Mol Sci. 2018;19:270-287.

[73] Baiyee B, Ito SI, Sunpapao A. *Trichoderma asperellum* T1 mediated antifungal activity and induced defense response against leaf spot fungi in lettuce (*Lactuca sativa* L.). Physiol Mol Plant Pathol. 2019;106:96-101.

[74] Unger C., Wilhelm I., Jünger R., Thalmann R. (2006) Evidence of induced resistance of tomato plants against *Phytophthora infestans* by a water extract of dried biomass of *Penicillium chrysogenum*. Journal of Plant Diseases and Protection, 113 (5), 225-233, 2006, ISSN 1861-3829.

[75] El-Sayed SA, Shad Ali G. *Aspergillus flavipes* is a novel efficient biocontrol agent of *Phytophthora parasitica*. Biological Control. 2019;10:4072-4093. DOI: 10.1016/j.biocontrol.2019.104072

[76] Samuels GJ, Hebbar PK. *Trichoderma*: Identification and Agricultural Applications. 2nd ed. APS Press: Boston. 2015. 196 p.

[77] Ezziyyani M, Sánchez CP, Ahmed AS, Requena ME, Castillo MEC. *Trichoderma harzianum* como biofungicida para el biocontrol de *Phytophthora capsici* en plantas de pimiento (*Capsicum annuum* L.). Anales de biología. 2004;26:35-45.

[78] Berger G, Czarnocka K, Cochard B, Oszako T, Lefort F. Biocontrol endotherapy with *Trichoderma* spp. and *Bacillus amyloliquefaciens* against *Phytophthora* spp.: A comparative study with phosphite treatment on *Quercus robur* and *Fagus sylvatica*. J Agri Sci Technol A. 2015;5:428-439. DOI: 10.17265/2161-6256/2015.06.005

[79] Tsahouridou PC, Thanassoulopoulos CC. Proliferation of *Trichoderma koningii* in the tomato rhizosphere and the suppression of damping-off by *Sclerotium rolfsii*. Soil Bio Biochem. 2002;34(6):767-776.

[80] Nusaibah A, Musa H. A review report on the mechanism of *Trichoderma* spp. as biological control agent of the basal stem rot (BSR) disease of *Elaeis guineensis*. In: Shah MM, editor. Hanbook of *Trichoderma* - The Most Widely Used Fungicide. Intechopen; 2019. DOI: 10.5772/intechopen.84469

[81] Elsherbiny EA, Amin BH, Aleem B, Kingsley KL, Bennett, J. W. *Trichoderma* Volatile Organic Compounds as a Biofumigation Tool against Late Blight Pathogen *Phytophthora infestans* in Postharvest Potato Tubers. J Agri Food Chem. 2020;68(31):8163-8171. DOI: 10.1021/acs.jafc.0c03150

[82] El-Hasan A, Walker F, Schöne J, Buchenauer H. Detection of viridiofungin A and other antifungal metabolites excreted by *Trichoderma harzianum* active against different plant pathogens. European J Plant Pathol. 2009;124(3):457-470. DOI: 10.1007/s10658-009-9433-3

[83] Hossain MM, Sultana F, Kubota M, Koyama H, Hyakumachi M. The plant growth-promoting fungus *Penicillium simplicissimum* GP17-2 induces resistance in *Arabidopsis thaliana* by

activation of multiple defense signals. Plant Cell Physiol. 2007;48:1724-1736.

[84] Khan SA, Hamayun M, Yoon H, Kim HY, Suh J, Hwang SK, Kim JM, Lee IJ, Choo YS, Yoon UH, Kong WS. Plant growth promotion and *Penicillium citrinum*. BMC Microbiol. 2008;8:1-10.

[85] Elsharkawy MM, Shimizu M, Takahashi H, Hyakumachi M. Induction of systemic resistance against Cucumber mosaic virus by *Penicillium simplicissimum* GP17-2 in Arabidopsis and tobacco. Plant Pathol. 2012;61:964-976.

[86] Srinivasan R, Prabhu G, Prasad M, Mishra M, Chaudhary M, Srivastava R. *Penicillium*. In: Amaresan Senthil N, Kumar M, Annapurna K, Kumar K, Sankaranarayanan A, editors. Handbook of Beneficial Microbes in Agro-Ecology, Bacteria and Fungi. 2nd ed. ACADEMIC Press. 2020;16:651-667. DOI: 10.1016/b978-0-12-823414-3.00032-0

[87] Myrchiang P, Dkhar S, Devi R. Studies on endophytic fungi associated with medicinally important aromatic plant *Artemisia nilagirica* (C.B. Clarke) Pamp., and their antagonistic activity against *Phytophthora infestans*. J Advanced Lab Res Bio. 2014;5(4): 112-119.

[88] Ma Y, Chang Z, Zhao J, Zhou M. Antifungal activity of *Penicillium striatisporum* Pst10 and its biocontrol effect on *Phytophthora* root rot of chilli pepper. Biological Control. 2008;44(1):24-31. DOI:10.1016/j.biocontrol.2007.10.005

[89] Nagaoka T, Nakata K, Kouno K, Ando T. Antifungal activity of oosporein from an antagonistic fungus against *Phytophthora infestans*. Z Naturforsch C J Biosci. 2004;59(4):302-304. DOI: 10.1515/znc-2004-3-432.

[90] De Vries S, Von Dahlen JK, Schnake A, Ginschel S, Schulz B, Rose LE. Broad-spectrum inhibition of *Phytophthora infestans* by fungal endophytes. FEMS Microbiol Ecol. 2018;94(4):37-53. DOI: 10.1093/femsec/fiy037.

[91] Morrison K, Arseneault T, Novinscak A, Filion M. Phenazine-1-Carboxylic Acid production by *Pseudomonas fluorescens* LBUM636 alters *Phytophthora infestans* growth and late blight development. Amer Phytopathol Society. 2017; 107: 273-279. DOI: 10.1094/phyto-06-16-0247-r

[92] Nagaokaa T, Nakatab K, Kounoa K, Andoa T. Antifungal activity of oosporein from an antagonistic fungus against *Phytophthora infestans.* verlag der zeitschrift für naturforschung. J Biosci. 2004;59:302-304. DOI: 10.1515/znc-2004-3-432

[93] Shuji T, Nishio N, Kai K, Hagiwara D, Ogata Y, Tojo M, Jun-ichi S, Judelson H, Kawaguchi T. Chemical genetic approach using β-rubromycin reveals that a RIO kinase-like protein is involved in morphological development in *Phytophthora infestans*. Scientific reports. 2020; 10: 223-226. DOI: 10.1038/s41598-020-79326-7

[94] Youyou W, Zhang C, Liang J, Wu L, Gao W, Jiang J. Iturin A Extracted From *Bacillus subtilis* WL-2 Affects *Phytophthora infestans* via Cell Structure Disruption, Oxidative Stress, and Energy Supply Dysfunction. Front Microbiol. 2020;11:(53)60-83. DOI: 10.3389/fmicb.2020.536083

[95] Youyou W, Liang J, Zhang C, Wang L, Gao W, Jizhi J. *Bacillus megaterium* WL-3 lipopeptides collaborate against *Phytophthora infestans* to control potato late blight and promote potato plant growth. Front Microbiol. 2020;11:16-32. DOI: 10.3389/fmicb.2020.01602

[96] Yeop JL, Surk-Sik M, Bong SY, Dong YI, Kook HB. Thiobutacin, a Novel Antifungal and Antioomycete Antibiotic from *Lechevalieria aerocolonigenes*. J Nat Prod. 2004;67:2076-2078. DOI: 10.1021/np049786v

[97] Son SW, Kim HY, Choi GJ, Lim HK, Jang KS, Lee SO, Lee S, Sung ND, Kim JC. Bikaverin and fusaric acid from *Fusarium oxysporum* show antioomycete activity against *Phytophthora infestans*. App Microbiol, Journal of Applied Microbiology. 2008;104:692-698. DO: I10.1111/j.1365-2672.2007.03581.x

[98] Puopolo G, Cimmino A, Palmieri MC, Giovannini O, Evidente A, Pertot I. *Lysobacter capsici* AZ78 produces cyclo(L-Pro-L-Tyr), a 2,5-diketopiperazine with toxic activity against sporangia of *Phytophthora infestans* and *Plasmopara viticola*. J App Microbio. 2014;117:1168-1180. DOI: 10.1111/jam.12611

[99] Yang X, Qiu D, Yang H, Liu Z, Zeng H, Yuan J. Antifungal activity of xenocoumacin 1 from *Xenorhabdus nematophilus* var. pekingensis against *Phytophthora infestans*. W J Microbiol Biotechnol. 2011;27:523-528. DOI: 10.1007/s11274-010-0485-5

[100] Mizushina Y, Ueno T, Oda M, Yamaguchi T, Saneyoshi M, Sakaguchi K. The biochemical mode of inhibition of DNA polymerase L by α-rubromycin. Biochim Biophys Acta. 2000;15(23): 172-181. DOI: 10.1016/S0304-4165(00)00119-7

[101] Shafi J, Tian H, Mingshan J. *Bacillus* species as versatile weapons for plant pathogens: a review. Biotech Biotechnol Equip, 2017;31:446-459. DOI: 10.1080/13102818.2017.1286950

[102] Druzhinina IS, Seidl-Seiboth V, Herrera-Estrella A, Horwitz AB, Kenerley MC, Monte E, Mukherjee PK, Zeilinger S, Grigoriev IV, Kubicek CP. *Trichoderma*: the genomics of opportunistic success. Nat Rev Microbio.2011;9:749-759.

[103] Harman GE. Trichoderma-not just for biocontrol anymore. Phytoparasitica. 2011;39:103-108.

[104] Horner NR, Grenville-Briggs LJ, West P. The oomycete *Pythium oligandrum* expresses putative effectors during mycoparasitism of *Phytophthora infestans* and is amenable to transformation. Fungal Bio. 2012;116:24-41.

[105] Liang D, Andersen CB, Vetukuri RR, Dou D, Grenville-Briggs LJ. Horizontal gene transfer and tandem duplication shape the unique co-enzyme complement of the Mycoparasitic oomycetes *Pythium oligandrum* and *Pythium periplocum*. Front Microbio. 2020;11:581-598.

[106] Le Floch G, Rey P, Déniel F, Benhamou N, Picard K, Tirilly Y. Enhancement of development and induction of resistance in tomato plants by the antagonist, *Pythium oligandrum*. Agronomie. 2003;23:455-460.

[107] Takenaka S, Yamaguchi K, Masunaka A, Hase S, Inoue T, Takahashi H. Implications of oligomeric forms of POD-1 and POD-2 proteins isolated from cell walls of the biocontrol agent *Pythium oligandrum* in relation to their ability to induce defense reactions in tomato. J Plant Physiol. 2011;168:1972-1979.

[108] Ouyang Z, Li X, Huang L, Hong Y, Zhang Y, Zhang H, Li D, Song F.Elicitin-like proteins Oli-D1 and Oli-D2 from *Pythium oligandrum* trigger hypersensitive response in *Nicotiana benthamiana* and induce resistance against *Botrytis cinerea* in tomato. Mol Plant Pathol. 2015;16:238-250.

[109] Lorito M, Woo SL, Harman GE, Monte E. Translational research on

Trichoderma: From 'omics' to the field. Ann Rev Phytopathol. 2010;48:395-417.

[110] Druzhinina IS, Seidl-Seiboth V, Herrera-Estrella A, Horwitz BA, Kenerley CM, Monte E. *Trichoderma*: The genomics of opportunistic success. Nat Rev Microbiol. 2011;9:749-759.

[111] Mukherjee PK, Buensanteai N, Morán-Diez ME, Druzhinina I, Kenerley CM. Functional analysis of non-ribosomal peptide synthetases (NRPSs) in *Trichoderma virens* reveals a polyketide synthase (PKS)/NRPS hybrid enzyme involved in the induced systemic resistance response in maize. Microbiol. 2012;158:155-165.

[112] Benítez T, Rincón A, Carmen Limón M. Biocontrol mechanisms of *Trichoderma* strains. International Microbiology. 2004;7:249-260.

[113] De Vries S, Von Dahlen JK, Schnake A, Ginschel S, Schulz B, Rose LE. Broad-spectrum inhibition of *Phytophthora infestans* by fungal endophytes. FEMS Microbiol Eco. 2018;94:1-15.

[114] Franceschi VR, Grimes HD. Induction of soybean vegetative storage proteins and anthocyanins by low-level atmospheric methyl jasmonate. Proc Natl Acad Sci USA 1991;88:6745-6749.

[115] Feys BJ, Benedetti CE, Penfold CN. Arabidopsis mutants selected for resistance to the phytotoxin coronatine are male sterile, insensitive to methyl jasmonate and resistant to a bacterial pathogen. Plant Cell. 1994;6:751-759.

[116] Li J, Brader G, Kariola T. WRKY70 modulates the selection of signaling pathways in plant defense. Plant J. 2006;46:477-491.

[117] Shan X, Zhang Y, Peng W. Molecular mechanism for jasmonate-induction of anthocyanin accumulation in Arabidopsis. J Exp Bot. 2009;60: 3849-3860.

[118] Cohen J, Gisi U, Niderman T. Local and system protection against *phytophthora infestans* induced in potato and tomato plants by jasmonic acid and jasmonic methyl ester. Phytopathol. 1993;83:1054-1062.

[119] Halim VA, Altmann S, Ellinger D, Eschen-Lippold L, Miersch O, Scheel D, Rosahl S. PAMP-induced defense responses in potato require both salicylic acid and jasmonic acid. Plant J. 2009;57:230-242.

[120] Vinale F, Sivassithamparam K, Ghisalbert EL, Marra R, Woo SL, Larito M. *Trichoderma*-plant-pathogen interactions. Soil Bio Biochem. 2008;40:1-10.

[121] Miethke M. Molecular strategies of microbial iron assimilation from high-affinity complexes to cofactor assembly systems. Metallomics. 2013;5:15-28.

[122] Mukherjee PK, Buensanteai N, Morán-Diez ME, Druzhinina I, Kenerley CM. Functional analysis of non-ribosomal peptide synthetases (NRPSs) in *Trichoderma virens* reveals a polyketide synthase (PKS)/NRPS hybrid enzyme involved in the induced systemic resistance response in maize. Microbiology. 2012;158:155-165.

[123] Lehner SM, Atanasova L, Neumann NK, Krska R, Lemmens M. Isotope-assisted screening for iron-containing metabolites reveals high diversity among known and unknown siderophores produced by *Trichoderma* spp. App Envir Microbiol. 2013;79:18-31.

[124] Si-Ammour A, Mauch-Mani B, Mauch F. Quantification of induced resistance against *Phytophthora* species expressing GFP as a vital marker: beta-aminobutyric acid but not BTH

protects potato and Arabidopsis from infection. Mol Plant Pathol. 2003;4:237-248.

[125] Spoel SH, Dong X. How do plants achieve immunity? Defence without specialized immune cells. Nat Rev Immunol. 2012;12:89-100.

[126] Pieterse CM, Zamioudis C, Berendsen R L, Weller DM, Van Wees SC, Bakker PA. Induced systemic resistance by beneficial microbes. Annu Rev Phytopathol. 2014;52: 347-375.

[127] Halim VA, Eschen-Lippold L, Altmann S, Birschwilks M, Scheel D, Rosahl S. Salicylic Acid Is Important for Basal Defense of *Solanum tuberosum* Against *Phytophthora infestans*. MPMI. 2007;20:1346-1352.

[128] Sharma K, Bruns C, Butz AF, Finckh MR. Effects of fertilizers and plant strengtheners on the susceptibility of tomatoes to single and mixed isolates of *Phytophthora infestans*. European Journal of Plant Pathology. 2012;133:739-751.

[129] Kumar R, Biswas SK, Kumar V, Lal K, Kumar U, Chowdary VT. Synthesis of defense enzymes in potato in induced resistance against late blight using inorganic chemicals as inducer. Plant Pathol J. 2017;16:130-137.

[130] Li J, Zhu L, Lu G, Zhan XB, Lin CC, Zheng ZY. Curdlan b-1,3 Glucooligosaccharides induce the defense responses against *Phytophthora infestans* infection of potato (*Solanum tuberosum* L. cv. McCain G1) leaf cells. Pols One. 2014;9:e97197.

[131] Conrath U, Linke C, Jeblick W, Geigenberger P, Quick W, Neuhaus H. Enhanced resistance to Phytophthora infestans and Alternaria solani in leaves and tubers, respectively, of potato plants with decreased activity of the plastidic ATP/ADP transporter. Planta. 2003;217:(1)75-83. http://www.jstor.org/stable/23387804.

Chapter 4

Endophytic Microorganisms as an Alternative for the Biocontrol of *Phytophthora* spp.

Hernando José Bolivar-Anillo, Victoria E. González-Rodríguez, Giovanna Reyes Almeida, Inmaculada Izquierdo-Bueno, Javier Moraga, María Carbú, Jesús M. Cantoral and Carlos Garrido

Abstract

The genus *Phytophthora* with more than 100 described species and 58 officially recognized, phylogenetically distributed in ten clades, are important pathogenic oomycete chromists that cause important diseases in agricultural crops, trees and forests worldwide. This genus is known as "The Plant Destroyer" which causes great economic losses with costs between 2 and 7 billion dollars per year in agricultural systems and unquantifiable losses in natural ecosystems. The host plants of the genus *Phytophthora* can vary from a wide range in some species to only one host, however, the host plants of the new species are still being determined and therefore the range continues to expand, that makes control exceedingly difficult. Plant damage can range from alterations in roots, fruits, trunks, stems, foliage and crown to invasive processes in highly susceptible species. Considering the wide range of hosts and organs that can be affected by *Phytophthora*, the use of endophytic microorganisms for the biocontrol of this phytopathogen can be an alternative to avoid losses of both crops and forests worldwide. Endophytes are microorganisms that live inside plant tissues without causing disease under any circumstances. The fact that endophytic microorganisms are able to colonize an ecological niche similar to that of some plant pathogens qualifies them as potential biocontrol agents. This chapter describes the endophytic bacteria and fungi isolated from different plant species that have shown antagonistic activity against different species of *Phytophthora*, as well as the metabolites isolated from these microorganisms that have shown fungicide activity and other biocontrol strategies (enzyme production, siderophores, substrate competition, among others) against *Phytophthora*.

Keywords: biological control agents, biocontrol, inhibitory mechanims, endophytic fungus, bacteria

1. Introduction

Phytopathogenic microorganisms are one of the main factors to causes losses in yield and quality of the crop along the world (worldwide). The economic losses due to diseases caused by microorganisms during pre- and post-harvest has been estimated to be between 30–40%, reaching almost 40 billion dollars worldwide annually [1, 2].

There is a great biodiversity of microorganisms that can cause diseases in plants. The group formed by phytopathogenic oomycetes (fungal-like organisms), is one of the most important and oldest. They have affected humankind since the beginning of agriculture in early civilisations [3]. During the last few centuries, these pathogens were responsible for the Potato Famine in Ireland, also known as the Great Famine, which caused almost a million deaths and triggered a mass migration in 1840 in that country [4, 5]. Even today, *Phytophthora infestans* is the causal agent of this disease in potatoes, it being the most important biotic limitation for the production of this tuber worldwide [4]. Other species of oomycetes, such as *Phytophthora ramorum*, do not only affect agriculture but also the environment, as they cause several diseases in many species of trees. As a consequence of the loss of forest mass due to infections and dead plants, it has been estimated an indirect impact on the environment that could reach a cumulative loss of 230–580 megatons of dissolved CO_2 during the last century [3]. Currently, these phytopathogens continue to represent a significant danger in agricultural and forestry systems because they have accelerated their evolution. This is caused by the continued use of fungicides, together with dispersal dependent on anthropogenic activities and climate (i.e. natural aerial dispersal and climate change). The use of monocultures as well as the greater use of perennial crops also increase the sexual recombination events of the populations of these oomycetes [4, 6]. This could cause an adaptation and improvement of these pathogens that would allow them to expand the range of hosts [4].

Among the phytopathogenic oomycetes, those of the genus *Phytophthora* are the best studied [1]. The genus *Phytophthora* is presently placed in the kingdom *Straminipila*, phylum *Heterokonta*, sub-phylum *Pernosporomycotina*, class *Pernosporomycetes* (Oomycetes), subclass *Pernosporomycetidae*, order *Pythiales* and family *Pythiaceae* [7]. *Phytophthora* has more than 100 described species and 58 officially recognized, phylogenetically distributed in ten clades, are usually soil-borne plant pathogens that cause important diseases in agricultural crops, trees and forests worldwide [8, 9]. This pathogen can present biotrophic, necrotrophic, or hemibiotrophic lifestyles [1, 3]. They reproduce asexually giving rise to sporangia, which divide into zoospores. When conditions are favourable, zoospores germinate to form mycelia or a specialized infection structure called appressorium. Sporangia can also germinate directly to produce mycelia or form an appressorium. Both sporangia and zoospores are important cells in the dissemination and infection processes [1].

Among the crops that can be infected by the genus *Phytophthora* are potato, tobacco, soybean, avocado, macadamia, cocoa, rice, tomato, pistachio, red pepper, strawberry, raspberry, among others [9–12]. Natural vegetation and ornamentals can also be infected by *Phytophthora* species, i.e. oaks, alder, holm, chestnut, cork oak, beech, rhododendron, viburnum, magnolia, pieris, among others [11–14]. Some species are highly specific to the host (i.e. *P. sojae*) or with a wide range of possible hosts (i.e. *P. cinnamomi*). However, the host plants of the new species are still being determined and, therefore, the range continues to expand, making control exceedingly difficult [7, 11]. Plant damage can range from alterations in roots, fruits, trunks, stems, foliage and crown to invasive processes in highly susceptible species [8, 9].

The control of infections caused by *Phytophthora*, in agriculture, forestry and natural systems is very limited. The fungicides available are usually not efficient against oomycetes since they are not true fungi [11]. Furthermore, the use of chemical fungicides is being increasingly restricted due to the adverse effects they produce on human, animal and environmental health [15]. As an alternative to the use of chemical products, the idea of using antagonistic microorganisms or the metabolites that they produce is proposed for the biocontrol of these oomycetes. The microorganisms used for biocontrol do not have negative effects on human or animal health and are considered friendly with the environment. Biocontrol carried out by microorganisms offers multiple modes of action, both direct, indirect or mixed, in addition, it prevents the appearance of resistance, which makes them an attractive alternative or complement for the control of phytopathogens [16]. The ability to biocontrol diseases through the use of microorganisms highlights the importance of interactions between the plant, the pathogen, the antagonist, the microbial community associated with the plant and environmental conditions [17, 18]. In this sense, most of the microorganisms used in biological control have been isolated from areas related to plants such as the rhizosphere, endosphere, phyllosphere, spermosphere, among others [19]. Although rhizosphere microorganisms are the most used in biocontrol, in recent decades a considerable number of endophytic microorganisms have been studied for their ability to biocontrol and for being a new source of natural products for use in agriculture [18, 20–22]. Therefore, this chapter describes the endophytic bacteria and fungi isolated from different plant species that have shown antagonistic activity against different species of *Phytophthora*, as well as the metabolites isolated from these microorganisms that have shown fungicide activity and other biocontrol strategies (enzyme production, siderophores, substrate competition, among others) against *Phytophthora*.

2. Endophytic microorganisms as a biocontrol strategy

Endophytes are microorganisms that are found inside plant tissues during at least part of their life cycle. They do not cause disease under any circumstances, and many

Figure 1.
Mechanisms of biocontrol showed by endophytic microorganisms.

show properties that promote plant growth [23, 24]. Approximately 300,000 species of plants have been described, and it is believed that each may possess different genera and species of endophytic microorganisms. However, it has only been studied the endophytic microbiome of 1–2% of plants. There are many unexplored fields of research on endophytes and their potential as biocontrol agents [25–28]. Although most endophytes are considered commensals, a large number of them establish mutualistic relationships with their host plant, playing a fundamental role in the adaptation of plants to biotic and abiotic factors [29–32]. Their use as biocontrol agents is considered one of the main characteristics to be used in the control of phytopathogens in agriculture. In this way we could reduce or avoid the use of antimicrobial compounds of chemical origin [18]. Endophytes can exert their biocontrol activity through various mechanisms including competition for a niche or substrate, hyperparasitism, predation, allelochemical production (antibiotics, lytic enzymes, siderophores) and by inducing systemic resistance in plants (**Figure 1**) [26, 33]. Now, the efficiency of endophytes as biological control agents depends on factors such as the specificity of the host, the physical structure of the soil, environmental conditions, the growth phase and the physiological state of the plant, among others [18, 34]. The development of a disease in a plant by any phytopathogenic microorganism will depend on three factors: the plant-the microbiota-the pathogen, whose interaction will be influenced by environmental factors. The loss of balance in any of these three factors would therefore lead to the development of an infectious process or not. On the other hand, most endophytic microorganisms originate in the soil (rhizosphere), therefore their recruitment (by the plant) will depend on their existence in soil, which is because they are not always present [35].

3. Biocontrol of *Phytophthora* spp. by endophytic bacteria

The promotion of plant growth by endophytic bacteria can be carried out through direct or indirect mechanisms [26, 36]. Among the indirect mechanisms, there is the biological control of phytopathogens, which is carried out through various strategies such as competition for nutrients and space, antibiosis, production of lytic enzymes, inhibition of toxins and induction of defense mechanisms in plants. All these strategies can be compatible with each other and may co-act simultaneously or synergistically [16, 18, 26, 37]. In this regard, there have been various studies that have evaluated the potential of endophytic bacteria for the biocontrol of different species of *Phytophthora*. These bacteria have been isolated from different plant species, which has led to the identification of microorganisms and the mechanisms used by them to inhibit the growth of this oomycete. **Table 1** shows some endophytic bacteria isolated from different plant species and the possible mechanisms they use for the biocontrol of *Phytophthora* spp.

El-Sayed *et al.,* (2018) [38] isolated forty morphologically distinct bacterial from roots, stems and leaves of *Smilax bona-nox* L. and they belonged to the genera *Burkholderia, Pseudomonas, Xenophilus, Stenotrophomonas, Pantoea, Enterobactriaceae, Kosakonia, Microbacterium, Curtobacterium, Caulobacter, Lysinibacillus* and *Bacillus*. Out of these isolates, the ones that showed the highest *in vitro* growth inhibition capacity of 5 species of *Phytophthora* (*P. parasitica, P. cinnamomi, P. palmivora, P. tropicalis* and *P. capsici*) were two strains of *Pseudomonas fluorescences* (EA6 and EA14). The percentage of inhibition of mycelial growth against different strains of *P. parasitica* was between 47% and 80%. On the other hand, the crude proteins (extracellular hydrolytic enzymes) obtained from *P. fluorescence* EA6 were able to inhibit the mycelial growth of *P. parasitica*. The analysis of these proteins revealed that they were glucanolytic enzymes (β-1,3 and β-1,4 glucanases) which act by

Microorganisms	Plant species	Inhibitory mechanisms	Ref.
Pseudomonas fluorescences	*Smilax bona-nox* L	Glucanolytic enzymes	[38]
Burkholderia spp.	*Huperzia serrata*	Siderophores	[39]
Acinetobacter calcoaceticus	*Glycine max L.*	Siderophores	[40]
Bacillus cereus	*Lycopersicon esculentum*	Triggering the plant immune defense	[41]
Bacillus *Paenibacillus* *Lactococcus* *Pediococcus* *Enterobacteriaceae* *Cronobacter* *Pantoea*	Seeds Cucurbits	Antibiosis VOCs RNase activity	[42]
Streptomyces Microbispora	*Lens esculentus* *Cicer arietinum* L. *Pisum sativum* *Vicia faba* *Triticum vulgare*	Antibiosis Siderophores	[43]
Bacillus thuringiensis *B. vallismortis* *B. amyloliquefaciens*	*Cornus florida* *Carica papaya*	Antibiosis Triggering the plant immune defense	[44]
Pseudomonas putida	*Piper nigrum*	VOCs	[45]
Streptomyces deccanensis *Bacillus* spp. *Rhizobium radiobacter* *Pantoea dispersa* *Bacillus velezensis* *Acinetobacter* spp.	*Piper colubrinum*	Competition Antibiosis Triggering the plant immune defense	[46]
Streptomyces alboniger Pseudomonas taiwanensis *P. geniculata* *Enterobacter hormaechei* *Bacillus tequilensis* *B. flexus* *Arthrobacter phenanthrenivorans* *Delftia lacustris*	*Dodonaea viscosa* *Fagonia indica* *Caralluma tuberculata* *Calendula arvensis*	Antibiosis VOCs Cell wall degrading enzymes Siderophores	[47]
Bacillus megaterium	*Piper nigrum*	VOCs HCN Hydrolytic activity Siderophore	[48]
Pseudomonas aeruginosa *Chryseobacterium proteolyticum*	*Theobroma cacao*	VOCs Hydrolytic activity Siderophore HCN	[49]
Bacillus velezensis	*Olea europaea*	Antibiosis VOCs Cell wall degrading enzymes Siderophores	[50]
Alcaligenes spp.	*Hevea brasiliensis*	PCA	[51]
Bacillus siamensis *B. amyloliquefaciens* *B. velezenis* *B. methylotrophiycus*	*Piper nigrum*	Cell wall degrading enzymes Antibiosis	[52]

Table 1.
Endophytic bacteria able to biologically control Phytophthora spp.

hydrolyzing the cell wall of *Phytophthora*. In addition, the crude glucanolytic extract was shown to have higher activity than the purified β-1,3-glucanase enzyme, which means that these enzymes act synergistically on the cell wall of *Phytophthora*. Want *et al.,* (2010) [39] from *Huperzia serrata*, isolated the endophytic bacteria identified as *Burkholderia* spp. H-6, which was able to inhibit the *in vitro* mycelial growth of *Phytophthora capsici* with a diameter of inhibition zones of 23 mm. Furthermore, in greenhouse pot experiments, the soils treated with *Burkholderia* spp. densities of 10^6, 10^8 and 10^{10} CFU ml^{-1} reduced *P. capsici* infection in pepper seedlings by 51.7, 58.7 and 60.2%, respectively. This strain presented the ability to synthesize siderophores, which could be related to its biocontrol capacity. Zhao *et al.,* (2018) [40] isolated a total of 276 endophytic bacteria from *Glycine max* L. nodules, of which 6 had an inhibition capacity greater than 63% against *Phytophthora sojae* and were identified as *Bacillus cereus*, *Acinetobacter calcoaceticus, Enterobacter cloacae, Bacillus amyloliquefaciens, Pseudomonas putida* and *Ochrobactrum haematophilum.* The strain identified as *Acinetobacter calcoaceticus* DD16 was the one that presented the highest inhibition of mycelial growth of *P. sojae* with 71.14%. *A. calcoaceticus* DD16 caused morphological abnormal changes of fungal mycelia (e.g. lysis, formation of a protoplast ball at the end of hyphae, and split ends) that could be related to the production of anti-fungal substances and fungal cell-lysing enzymes. In addition, *A. calcoaceticus* DD16 was the strain that presented the highest capacity to produce siderophores (54.33 ± 0.093 μg mL^{-1}) and was capable of fixing nitrogen and producing indole acetic acid, activities related to the promotion of plant growth. The regression analysis showed a significant positive correlation between siderophore production and inhibition ratio against *P. sojae*. Melnick *et al.,* (2008) [41] isolated from *Lycopersicon esculentum* a strain of endophytic bacteria identified as *Bacillus cereus* BT8, which *in vitro* test did not show the ability to inhibit the mycelial growth of *Phytophthora capsici*. However, this strain exhibited the ability to colonize *Theobroma cacao* seedlings and reduce the severity of *Phytophthora capsici* infection. The suppression of *P. capsici* was only observed in leaves which were not inoculated with the endophytic bacteria after colonization of the plant in other leaves, which suggests that the mechanism of suppression of the disease is through the induction of defense mechanisms in the plants (Induced Systemic Resistance) rather than antagonistic mechanisms. Khalaf *et al.,* (2018) [42] isolated a total of 169 bacterial endophytes from seeds of diverse cultivated cucurbits (*Luffa acutangula*, *Curcubita moschata*, *Curcubita pepo*, *Lagenaria siceraria*, *Citrullus lanatus*, *Cucumis melo* and *Cucumis sativos*), of which 26% (44/169) of isolates showed anti-pathogenic traits *in vitro* against *Phytophthora capsici*, of these 44 isolates, 16 were obtained from *Cucumis melo* seeds. These bacteria with activity against *P. capsici* belonged to the genera *Bacillus*, *Paenibacillus*, *Lactococcus*, *Pediococcus*, *Enterobacteriaceae*, *Cronobacter* and *Pantoea*. Of these microorganisms, those of the genus *Bacillus*, *Paenibacillus*, *Enterobacteriaceae* and *Pantoea* showed acetoin/diacetyl production (volatile organic compounds VOCs) and RNase activity *in vitro*, known to be implicated in triggering the plant immune defense. Therefore, these bacteria may control the phytopathogen directly (antibiosis) and/or indirectly (induction of host defense).

Misk and Franco (2011) [43] isolated thirty-six actinobacterial strains from different plants (root, stem and leaf), lentil (*Lens esculentus*), chickpea (*Cicer arietinum* L.), pea (*Pisum sativum*), faba bean (*Vicia faba*) and wheat (*Triticum vulgare*). Eleven of the isolates had antimicrobial activity against *Phytophthora medicaginis*, where ten of those isolates belonged to *Streptomyces* and one to *Microbispora*. The strains identified as *Streptomyces* spp. WRA1 and BSA25 were the most efficient as they significantly inhibited 100% and 85% *in vitro* of *P. medicaginis*, respectively, which showed a good capacity to produce siderophores. Furthermore, *in vivo* tests both strains (WRA1 and BSA25) significantly inhibited

P. medicaginis root rot compared to infected control. This inhibition capacity against *P. medicaginis* could be related to their antibiotic and siderophores production. Bhusal and Mmbaga (2020) [44] evaluated the biocontrol capacity of three endophytic bacterias *Bacillus thuringiensis* isolated from flowering dogwood stem; *B. vallismortis*; and *B. amyloliquefaciens* isolated from papaya stem against *Phytophthora capsici*. *B. amyloliquefaciens* was the most effective in suppressing *P. capsici* mycelial growth *in vitro* up to 46.62%, followed by *B. vallismortis* 45.95% and *B. thuringiensis* 27.59%. Under the greenhouse environment, *B. amyloliquefaciens* and *B. vallismortis* were most effective in suppressing *P. capsici* symptoms. Agisha *et al.*, (2019) [45] evaluated the antimicrobial capacity on phytopathogens of VOCs produced by the black pepper endophytic bacterium, *Pseudomonas putida*. Of the VOCs produced by *P. putida*, those identified as 2,5-dimethyl pyrazine; 2-methyl pyrazine; dimethyl trisulphide; 2-ethyl 5-methyl pyrazine; and 2-ethyl 3, 6-dimethyl pyrazine showed inhibitory activity (sealed plate method) against *Phytophthora capsici*. Among these VOCs, 2-ethyl-3, 6-dimethyl pyrazine was the most effective with an EC50, EC90 and EC95 of 66.1 $\mu g\ cm^{-3}$, 244.8 $\mu g\ cm^{-3}$ and 382.1 $\mu g\ cm^{-3}$, respectively. In trials to evaluate the effect of VOCs against *Phytophthora* rot on black pepper shoot cuttings, 2, 5 dimethyl pyrazine, 2-ethyl 5-methyl pyrazine and 2-ethyl 3, 6-dimethyl pyrazine displayed reduction of lesion at 21 $\mu g\ cm^{-3}$ and, 2-methyl pyrazine at 42 $\mu g\ cm^{-3}$ with no signs of toxicity. While in the tests for fumigant activity of volatiles, dimethyl trisulphide demonstrated complete inhibition against *P. capsici* at a concentration of 6.25 $\mu g\ cm^{-3}$, which demonstrated that these VOCs can be an alternative for the control of *P. capsici* infections. Kollakkodan *et al.*, (2020) [46] isolated endophytic bacteria from the roots, stem and leaves of *Piper colubrinum*. Seven of these isolates showed *in vitro* inhibition capacity against *Phytophthora capsici* with zones of inhibition between 2.4 and 5.8 mm, which were identified as *Streptomyces deccanensis*, *Bacillus* spp., *Rhizobium radiobacter*, *Pantoea dispersa*, *Bacillus velezensis* (PCSE8), *Bacillus velezensis* (PCSE10) and *Acinetobacter* spp. The maximum inhibition zone was produced by the two strains of *B. velezensis*. In leaf assay (leaves of black pepper), the highest suppression of the disease was presented by the strains identified as *Pantoea dispersa* and *Bacillus velezensis* (PCSE10), with percentages of 74% and 79%, respectively. The mechanisms of these endophytic bacteria which are responsible for the inhibition of *P. capsica* seem to be mainly related to competition, antibiosis and triggering of the plant's immune defence. Iqrar *et al.*, (2021) [47] isolated endophytic bacteria from medicinal plants, *Dodonaea viscosa*, *Fagonia indica*, *Caralluma tuberculata* and *Calendula arvensis*. Bacteria that exhibited biocontrol activity on screening assays (production of cell wall degrading enzymes and siderophores) were identified as *Streptomyces alboniger*, *Pseudomonas taiwanensis*, *Pseudomonas geniculata*, *Enterobacter hormaechei*, *Bacillus pheustrivo*, *Bacillus flexus* and *Delftiartiabacteris*. In the *in vitro* growth inhibition test against *Phytophthora parasitica*, the highest inhibition was presented by the bacterium identified as *P. taiwanensis* with 55%, as well as in the bipartite split-plate growth inhibition assays (VOCs) with an inhibition of 80%. In addition, the crude extracts from the culture of this bacterium presented an inhibition of 92% at a concentration of 400 $\mu g\ mL^{-1}$ and the ethyl acetate extract presented an inhibition of 60%. The hyphae of *P. parasitica* subjected to these extracts showed alterations in their structure (convoluted, swollen nodes and abnormal growth of hyphae). The inhibition capacity of these endophytic bacteria on *P. parasitica* seems to be related to multiple mechanisms of action such as antibiosis, VOCs, cell wall degrading enzymes and siderophores. Munjal *et al.*, (2016) [48] isolated an endophytic bacterium identified as *Bacillus megaterium* from the black pepper root that was capable of inhibiting different phytopathogens *in vitro*, including *Phytophthora capsici*. This bacterium exhibited the ability to produce hydrogen cyanide (HCN), protease, cellulase and siderophore. In VOCs' activity tests, it was observed a growth inhibition of *P. capsica* of 28%. These VOCs were mainly composed of 2,5-dimethyl pyrazine, 2-ethyl-3-methyl pyrazine, 2-ethyl pyrazine and 2-methyl pyrazine and they were able to inhibit individual mycelial growth by more than 60% at a concentration of 336 μg mL^{-1}. Among these VOCs, the most effective was 2-ethyl-3-methyl pyrazine, which 100% inhibited the mycelial growth of *P. capsici* at a concentration of 168 μg mL^{-1}. Therefore, the antagonistic activity of this bacterium is related to the ability to produce VOCs, HCN, protease, cellulase and siderophore. Alsultan *et al.,* (2019) [49] isolated 103 endophytic bacteria from cacao plants (leaves, branches and fruits) of which two that showed an 80% *in vitro* inhibition of *P. palmivora* and were identified as *Pseudomonas aeruginosa* and *Chryseobacterium proteolyticum*. While in the culture filtrate test, the inhibition percentages were 100% and 62% to *P. aeruginosa* and *Ch. proteolyticum*, respectively. In the volatile metabolites test, *P. aeruginosa* and *C. proteolyticum* strains showed an inhibition of pathogen growth of 61.88% and 60.94%, respectively. The VOCs produced by *P. aeruginosa* were identified as eicosane, hexatriacontane, tetratetracontane, trans-2-decenoic acid and 1-phenanthrenecarboxylic acid, 1,2,3,4,4α,9,10,10α-octahydro-1,4α-dimethyl-7-(1-methylethyl), while those produced by *C. zproteolyticum* were identified as eicosane, tetratetracontane, heneicosane, hexatriacontane and phenol 2,4-bis(1,1-dimethylethyl). Regarding the hydrolytic activity, these two strains were capable of producing cellulase, protease, pectinase and lipase. Only *P. aeruginosa* was able to produce siderophores and HCN. The inhibition capacity of both strains is related to the capacity to produce hydrolytic enzymes, VOCs, HCN and siderophores that can act individually or synergistically. Cheffi *et al.,* (2019) [50] isolated the endophytic bacterium identified as *Bacillus velezensis* from olive trees, which exhibited an inhibition ranged from 40 to 75% with oomycetes, including *Phytophthora ramorum*, *P. cactorum*, *P. cryptogea*, *P. plurivora* and *P. rosacearum*. Regarding its biocontrol capacity, *B. velezensis* presented the capacity to produce VOCs, among which ethylbenzene, phenylethyl alcohol, E-caryophyllene and cyclo (Leu-Pro) were detected. Through genome analysis, diverse secondary metabolite clusters were uncovered such as bacillomycin, amylocyclin, mersacidin, bacilysin, macrolactin, bacillibactin, bacillaene, surfactin, fengycin, dicidin, subtilin and locillomycin. The analysis of the culture extracts by means of LC–MS, detected the production of surfactin B, surfactin C15, plipastatin B1, Fengycin B, IX and XII. Furthermore, this strain was able to produce cell wall degrading enzymes (protease, chitinase and glucanase) and siderophores. All these metabolites could be responsible for the inhibition capacity of *B. velezensis* on these oomycetes. Abraham *et al.,* (2015) [51] isolated the endophytic bacterium identified as *Alcaligenes* spp. from *Hevea brasiliensis,* that presents antagonistic activity against *Phytophthora meadii*. By means of the spectrometric study of the culture supernatant of *Alcaligenes* spp., it was established that the compound identified as phenazine-1-carboxylic acid showed inhibition of *P. meadii* growth. The minimum inhibitory concentration of this compound against *P. meadii* was optimized at 5 μg mL^{-1}. In addition, this compound presented zoospore-lytic activity, the structure of which was completely altered and lysis of the same occurred. The zoospores were not able to germinate when they were cultured in the presence of this compound. Ngo *et al.,* (2020) [52] isolated endophytic black pepper bacteria, of which six showed the ability to inhibit the growth of *Phytophthora* spp. by more than 60%. These bacteria were identified as *Bacillus siamensis*, *B. amyloliquefaciens*, *B. velezenis* and *B. methylotrophiycus*. These strains presented high chiti-nase and protease activities. In the *in vivo* test, the strains identified as *B. siamensis*, *B. velezensis* and *B. methylotrophycus* (EB.KN13) had the lowest rate of root disease (8.45–11.21%) and lower fatal rate (11.11–15.55%).

4. Biocontrol of *Phytophthora* spp. by endophytic fungi

Like bacteria, endophytic fungi can protect their host plant against both biotic and abiotic stressors; which are considered a rich source of bioactive metabolites [32, 53, 54]. Among the main mechanisms by which endophytic fungi prevent infections by phytopathogens are induced resistance, antibiosis, mycoparasitism, competition and extracellular enzymes [31, 32, 54]. **Table 2** summarizes the species of endophytic fungi with biocontrol capacity against *Phytophthora* spp. and the plant species from which they were isolated, revealing the wide diversity of endophytic fungi that can be used for the biocontrol of this phytopathogen. Hanada *et al.*, (2010) [55] evaluated the antagonistic capacity of endophytic fungi isolated from *Theobroma cacao* and *Theobroma grandiflorum* against *Phytophthora palmivora*.

Microorganisms	Plant species	Inhibitory mechanisms	Ref.
Trichoderma *Pestalotiopsis* *Curvularia* *Tolypocladium* *Fusarium*	*Theobroma cacao* *T. grandiflorum*	Antibiosis	[55]
Muscodor crispans	*Ananas ananassoides*	VOCs	[56]
Trichoderma viride *T. pseudokoningii*	*Piper nigrum*	Antibiosis	[57]
Trichoderma ovalisporum *T. theobromicola* *T. hamatum* *T. stilbohypoxyli* *T. caribbaeum var. aequatoriale* *T. theobromicola*	*Banisteriopsis caapi* *Theobroma cacao Theobroma gileri Theobroma cacao* *Theobroma gileri* *Cola praecuta*	Mycoparasitism Antibiosis Systemic induced resistance	[58]
Aureobasidium pullulans *Nigrospora oryzae* *Chaetomium globosum Trichoderma asperellum Penicillium commune*	*Espeletia* spp.	Antibiosis Competition for substrate	[59]
Phialocephala europaea	*Picea abies*	Antibiosis	[60]
Phoma terrestris *Fusarium oxysporum* *Ascomycete* spp.	*Panax quinquefolius*	Antibiosis Cell wall degrading enzymes	[61]
Cryptosporiopsis spp. *Phialocephala sphareoides*	*Picea abies*	Antibiosis	[62]
Ceriporia lacerate *Phomopsis* spp. *Diaporthe* spp. *Daldinia eschscholtzii* *Annulohypoxylon nitens* *Fusarium* spp.	*Piper nigrum*	Competition Antibiosis Mycoparasitism VOCs	[63]
Purpureocillium lilacinum	*Solanum lycopersicum*	Antibiosis Cell wall degrading enzymes	[64]
Xylaria spp.	*Haematoxylon brasiletto* Karst	Antibiosis VOCs	[65]
Hypoxylon anthochroum	*Gliricidia sepium*	Antibiosis	[66]

Table 2.
Endophytic fungi with biocontrol capacity against Phytophthora spp.

A total of 103 endophytic fungi were isolated of which ~70% showed some degree of reduction in the disease severity in three cacao pods. Eight isolates from genera *Trichoderma*, *Pestalotiopsis*, *Curvularia*, *Tolypocladium* and *Fusarium* showed the highest level of activity against the pathogen. The possible responsible mechanisms for the ability to inhibit *P. palmivora* were related to the production of bioactive compounds. Mitchell *et al.*, (2010) [56] evaluated the ability of the VOCs of the endophytic fungus *Muscodor crispans* isolated from *Ananas ananassoides* to inhibit the growth of phytopathogens, among which there were *Phytophthora cinnamomi* and *P. palmivora*. The VOCs produced by *M. crispans* that were composed mainly of propanoic acid, 2-methyl; propanoic acid, 2-methyl-; 1-butanol, 3-methyl-; 1-butanol, 3-methyl-, acetate; propanoic acid, 2-methyl-, 2-methylbutyl ester; and ethanol and were able to inhibit the growth of *Phytophthora cinnamomi* and *P. palmivora* by 100% with an IC_{50} ($\mu L\ mL^{-1}$) of 0.056 and < 0.02, respectively. Mathew *et al.*, (2011) [57] isolated two endophytic fungi identified as *Trichoderma viride* and *T. pseudokoningii* from black pepper plants which showed *in vitro* inhibition capacity against *Phytophthora capsici* with an inhibition percentage of 64.4% and 65.6%, respectively. In the *in vivo* study, the lowest percentage in the incidence and severity of the disease caused by *P. capsici* was presented by the strain identified as *T. viride*. Bae *et al.*, (2011) [58] evaluated the antagonism capacity against *Phytophthora capsici* of six species of *Trichoderma* (*T. ovalisporum, T. theobromicola, T. hamatum, T. stilbohypoxyli, T. caribbaeum* var. *aequatoriale* and *T. theobromicola*) isolated from *Banisteriopsis caapi*, *Theobroma cacao*, *Theobroma gileri*, and *Cola praecuta*. All strains except for *T. caribbaeum* var. *aequatoriale* showed the ability to parasitize the mycelium of *P. capsici*. However, the culture filters of *T. caribbaeum* var. *aequatoriale* completely prevented growth of *P. capsici*, while *T. stilbohypoxyli* and *T. ovalisporum* presented inhibition percentages of 56.5% and 30.7, respectively. In addition, it was shown that the inoculation of *Trichoderma* strains in pepper seedlings activated genes associated with responsive to stress. *In vivo* tests, the strain identified as *T. theobromicola* delayed the onset of disease symptoms for more than 3 days and between 26 and 60% of the pepper seedlings remained asymptomatic. Miles *et al.*, (2012) [59] studied the biocontrol potential of 100 fungal endophytes isolated from *Espeletia* spp. Among the phytopathogens used to measure this potential was *Phytophthora infestans*. The growth of *P. infestans in vitro* was completely inhibited by eight endophytes which were identified as *Aureobasidium pullulans*, *Nigrospora oryzae*, *Chaetomium globosum*, *Trichoderma asperellum* and *Penicillium commune*. The crude extract of the culture of *A. pullulans* and *P. commune* also showed the ability to inhibit 100% the growth of *P. infestans*. Tellenbach *et al.*, (2013) [60] evaluated the ability of *Phialocephala europaea* isolated from *Picea abies* to inhibit the growth of *Phytophthora citricola* s.l. The strain of *P. europaea* was able to reduce the growth of *P. citricola in vitro*. The four compounds isolated from this microorganism were identified as sclerin, sclerolide, sclerotinin A and sclerotinin B. Sclerin and sclerotinin A were the main compounds produced, which *in vitro* significantly reduced the growth of *P. citrícola* at a concentration of 30 $mg\ mL^{-1}$. Park *et al.*, (2015) [61] isolated the endophytic fungi identified as *Phoma terrestris*, *Fusarium oxysporum* and *Ascomycete* spp. from *Panax quinquefolius*, which inhibited the growth of *Phytophthora cactorum* with percentages between 64% to 82% and from 71% to 80% in the disk diffusion tests and fermentation broth tests, respectively. The main metabolites produced by *P. terrestris, F. oxysporum* and *Ascomycete* spp., were identified as N-amino-3-hydroxy-6-methoxyphthalimide, 3-methylthiobenzothiophene, phthalic acid, erucylamide and 2H-1-benzopyran-2-1, 3,4,5,6,7,8-hexahydro-4,7-dimethyl-. In the enzyme assays, the endophytic fungus identified as *P. terrestris* showed activity for the cellulase, xylanase, β-glucanase, pectinase and chitinase enzymes that could play a role in the inhibition of phytopathogens.

Terhonen *et al.,* (2016) [62] isolated the endophytic fungi identified as *Cryptosporiopsis* spp. and *Phialocephala sphareoides* from *Picea abies* which were able to inhibit the growth of *Phytophthora pini in vitro*. In addition, a decrease in the growth of *P. pini* was observed when the crude extract of the culture medium of *Cryptosporiopsis* spp. were tested. Subsequently, the analysis of the crude extract by UPLC-QTOF/MS was able to establish that the main metabolites produced by *Cryptosporiopsis* spp. had the following chemical formula $C_{19}H_{30}O_6$, $C_{20}H_{28}O_8$, $C_{20}H_{30}O_7$ and $C_{18}H_{28}O_6$. Sreeja *et al.,* (2016) [63] isolated 125 endophytic fungi from *Piper nigrum* which were evaluated to measure the ability to inhibit *Phytophthora capsici in vitro*. Of the 125 isolated fungi, 23 presented this capacity in more than 50%. The fungi with the highest inhibition capacity (78%) were identified as *Ceriporia lacerate*, *Phomopsis* spp. and *Diaporthe* spp. Other strains identified as *Daldinia eschscholtzii*, *Annulohypoxylon nitens* and *Fusarium* spp. presented inhibition capacity between 74% to 75%. Competition, VOCs antibiosis and mycoparasitism were reported to be among the biocontrol strategies for these fungi against *P. capsica*. Wang *et al.,* (2016) established by genome mining the biocontrol capacity of two strains of *Purpureocillium lilacinum* (PLBJ-1 and PLFJ-1) isolated from *Solanum lycopersicum*. Among the genes detected that may be useful in biocontrol were those that code for CAZymes, protease, glycoside hydrolases, and carbohydrate esterase. Regarding the production of secondary metabolites, genes coding for polyketide synthase, non-ribosomal peptide synthetase, terpene synthase and dimethylallyl tryptophan synthase were detected. Among these genes, those responsible for the synthesis of leucinostatin A and B were detected, which was confirmed by the production of mutants incapable of producing this compound. *In vitro* tests with the wild type and the mutant strain showed that the synthesis of leucinostatin A and B is closely related to the ability of these strains to inhibit the growth of *Phytophthora infestans* and *P. capsici*. Sanchez-Ortiz *et al.*, (2016) [65] evaluated the biocontrol capacity and VOCs of the endophytic fungus of *Haematoxylon brasiletto* Karst identified as *Xylaria* spp. PB3f3. The endophytic fungus was able to inhibit *Phytophthora capsici* by 48.3% *in vitro* and it was able to produce forty VOCs composed mainly of 3-methyl-1-butanol and thujopsene. Sánchez-Fernández *et al.* (2020) [66] studied antifungal and antioomycete activities of the compounds synthesized by the endophytic fungus *Hypoxylon anthochroum* isolated from *Gliricidia sepium*. The chemical study of the culture medium and the organic extracts of mycelium of the endophytic fungus led to the isolation of three isobenzofuranones: 7-hydroxy-4,6-dimethyl-3H-isobenzofuran-1-one **(1)**, 7-methoxy-4, 6-dimethyl-3H-isobenzofuran-1-one **(2)**, 6-formyl-4-methyl-7-methoxy-3H-isobenzofuran-1-one **(3)** and one compound was isolated for the first time as a natural product, 7- methoxy-4-methyl-3H-isobenzofuran-1-one **(4)** and another obtained by chemical synthesis 7-methoxy-6-methyl-3H-isobenzofuran-1-one **(5)**, which showed the ability to inhibit the radial growth of *Phytophthora capsici* with an IC_{50} mM of 0.76, 0.62,> 0.97,> 1.12 and 2.12 respectively. Regarding the ability to alter the permeability of the *P. capsici* membrane, compounds 1, 2 and 5 presented an IC_{50} mM of <1.40, 0.55 and 2.03, respectively. In addition, these compounds were able to inhibit the respiration of *P. capsici*, being 2 the most efficient with an IC_{50} mM of 0.34.

5. Conclusions

Currently, the control of infections caused by *Phytophthora* spp. is very complicated, mainly due to the fact that many of the fungicides available on the market are not effective against this oomycete and also many of them are associated with environmental and health damage. Therefore, the use of biocontrol agents as an alternative opens the possibility of using endophytic microorganisms, associated with the plant environment, which show great potential against this oomycete. Endophytic microorganisms isolated from different plant species have shown the ability to inhibit the growth of different *Phytophthora* species through various mechanisms such as antibiosis, VOCs, enzyme production, competition, among others. Therefore, the isolation of endophytic microorganisms and the study of their antagonistic capacity allows us to find new biocontrol agents, or their bioactive molecules, that allow controlling the enormous economic losses caused by *Phytophthora* spp.

Author details

Hernando José Bolivar-Anillo[1], Victoria E. González-Rodríguez[2], Giovanna Reyes Almeida[1], Inmaculada Izquierdo-Bueno[2], Javier Moraga[2], María Carbú[2], Jesús M. Cantoral[2] and Carlos Garrido[2]*

1 Programa de Microbiología, Facultad de Ciencias Básicas y Biomédicas, Universidad Simón Bolívar, Barranquilla, Colombia

2 Facultad de Ciencias del Mar y Ambientales, Laboratorio de Microbiología, Departamento de Biomedicina, Biotecnología y Salud Pública, Universidad de Cádiz, Puerto Real, España

*Address all correspondence to: carlos.garrido@uca.es

References

[1] Wang, T.; Gao, C.; Cheng, Y.; Li, Z.; Chen, J.; Guo, L.; Xu, J. Molecular diagnostics and detection of oomycetes on fiber crops. *Plants* **2020**, *9*, 1-22, doi:10.3390/plants9060769.

[2] Syed Ab Rahman, S.F.; Singh, E.; Pieterse, C.M.J.; Schenk, P.M. Emerging microbial biocontrol strategies for plant pathogens. *Plant Sci.* **2018**, *267*, 102-111, doi:10.1016/j.plantsci.2017.11.012.

[3] Covo, S. Genomic instability in fungal plant pathogens. *Genes (Basel).* **2020**, *11*, doi:10.3390/genes11040421.

[4] Corredor-Moreno, P.; Saunders, D.G.O. Expecting the unexpected: factors influencing the emergence of fungal and oomycete plant pathogens. *New Phytol.* **2020**, *225*, 118-125, doi:10.1111/nph.16007.

[5] Fones, H.N.; Bebber, D.P.; Chaloner, T.M.; Kay, W.T.; Steinberg, G.; Gurr, S.J. Threats to global food security from emerging fungal and oomycete crop pathogens. *Nat. Food* **2020**, *1*, 332-342, doi:10.1038/s43016-020-0075-0.

[6] Pandaranayaka, E.P.J.; Frenkel, O.; Elad, Y.; Prusky, D.; Harel, A. Network analysis exposes core functions in major lifestyles of fungal and oomycete plant pathogens. *BMC Genomics* **2019**, *20*, 1-15, doi:10.1186/s12864-019-6409-3.

[7] Roy, S.G.; Grünwald, N.J. The plant destroyer genus *Phytophthora* in the 21st century. *Rev. Plant Pathol.* **2014**, *6*, 387-412.

[8] Kroon, L.P.N.M.; Brouwer, H.; De Cock, A.W.A.M.; Govers, F. The genus *Phytophthora* anno 2012. *Phytopathology* **2012**, *102*, 348-364, doi:10.1094/PHYTO-01-11-0025.

[9] Gao, R.F.; Wang, J.Y.; Liu, K.W.; Yoshida, K.; Hsiao, Y.Y.; Shi, Y.X.; Tsai, K.C.; Chen, Y.Y.; Mitsuda, N.; Liang, C.K.; et al. Comparative analysis of *Phytophthora* genomes reveals oomycete pathogenesis in crops. *Heliyon* **2021**, *7*, e06317, doi:10.1016/j.heliyon.2021.e06317.

[10] de Andrade Lourenço, D.; Branco, I.; Choupina, A. Phytopathogenic oomycetes: a review focusing on *Phytophthora cinnamomi* and biotechnological approaches. *Mol. Biol. Rep.* **2020**, *47*, 9179-9188, doi:10.1007/s11033-020-05911-8.

[11] Hardham, A.R.; Blackman, L.M. *Phytophthora cinnamomi*. *Mol. Plant Pathol.* **2018**, *19*, 260-285, doi:10.1111/mpp.12568.

[12] Cline, E.T.; Farr, D.F.; Rossman, A.Y. A Synopsis of *Phytophthora* with Accurate Scientific Names, Host Range, and Geographic Distribution. *Plant Heal. Prog.* **2008**, *9*, 32, doi:10.1094/php-2008-0318-01-rv.

[13] Gibbs, J.N.; Lipscombe, M.A.; Peace, A.J. The impact of *Phytophthora* disease on riparian populations of common alder (*Alnus glutinosa*) in southern Britain. *Eur. J. For. Pathol.* **1999**, *29*, 39-50, doi:10.1046/j.1439-0329.1999.00129.x.

[14] Brasier, C.M.; Beales, P.A.; Kirk, S.A.; Denman, S.; Rose, J. *Phytophthora kernoviae* sp. nov., an invasive pathogen causing bleeding stem lesions on forest trees and foliar necrosis of ornamentals in the UK. *Mycol. Res.* **2005**, *109*, 853-859, doi:10.1017/S0953756205003357.

[15] Carbú, M.; González-Rodriguez, V. Garrido, C.; Husaini, C.; Cantoral, J. New biocontrol strategies for strawberry fungal pathogens. In *Strawberry: Growth, Development and Disease*; Amjad, H., Neri, D., Eds.; CABI: Boston, 2016; pp. 196-211.

[16] Bardin, M.; Ajouz, S.; Comby, M.; Lopez-Ferber, M.; Graillot, B.; Siegwart,

M.; Nicot, P.C. Is the efficacy of biological control against plant diseases likely to be more durable than that of chemical pesticides? *Front. Plant Sci.* **2015**, *6*, 1-14, doi:10.3389/fpls.2015.00566.

[17] Bosso, L.; Scelza, R.; Varlese, R.; Meca, G.; Testa, A.; Rao, M.A.; Cristinzio, G. Assessing the effectiveness of *Byssochlamys nivea* and *Scopulariopsis brumptii* in pentachlorophenol removal and biological control of two *Phytophthora* species. *Fungal Biol.* **2016**, *120*, 645-653, doi:10.1016/j.funbio.2016.01.004.

[18] Bolívar-Anillo, H.J.; Garrido, C.; Collado, I.G. Endophytic microorganisms for biocontrol of the phytopathogenic fungus *Botrytis cinerea*. *Phytochem. Rev.* **2019**, 1-20, doi:10.1007/s11101-019-09603-5.

[19] Lazarovits, G.; Turnbull, A.; Johnston-Monje, D. *Plant Health Management: Biological Control of Plant Pathogens*; Elsevier Ltd., 2014; Vol. 4; ISBN 9780080931395.

[20] Ryan, R.P.; Germaine, K.; Franks, A.; Ryan, D.J.; Dowling, D.N. Bacterial endophytes: Recent developments and applications. *FEMS Microbiol. Lett.* **2008**, *278*, 1-9, doi:10.1111/j.1574-6968.2007.00918.x.

[21] Eljounaidi, K.; Lee, S.K.; Bae, H. Bacterial endophytes as potential biocontrol agents of vascular wilt diseases – Review and future prospects. *Biol. Control* **2016**, *103*, 62-68, doi:10.1016/j.biocontrol.2016.07.013.

[22] Zheng, Y.K.; Qiao, X.G.; Miao, C.P.; Liu, K.; Chen, Y.W.; Xu, L.H.; Zhao, L.X. Diversity, distribution and biotechnological potential of endophytic fungi. *Ann. Microbiol.* **2016**, *66*, 529-542, doi:10.1007/s13213-015-1153-7.

[23] Le Cocq, K.; Gurr, S.J.; Hirsch, P.R.; Mauchline, T.H. Exploitation of endophytes for sustainable agricultural intensification. *Mol. Plant Pathol.* **2017**, *18*, 469-473, doi:10.1111/mpp.12483.

[24] Ludwig-Müller, J. Plants and endophytes: equal partners in secondary metabolite production? *Biotechnol. Lett.* **2015**, *37*, 1325-1334, doi:10.1007/s10529-015-1814-4.

[25] Aly, A.H.; Debbab, A.; Kjer, J.; Proksch, P. Fungal endophytes from higher plants: A prolific source of phytochemicals and other bioactive natural products. *Fungal Divers.* **2010**, *41*, 1-16, doi:10.1007/s13225-010-0034-4.

[26] Morales-Cedeño, L.R.; Orozco-Mosqueda, M. del C.; Loeza-Lara, P.D.; Parra-Cota, F.I.; de los Santos-Villalobos, S.; Santoyo, G. Plant growth-promoting bacterial endophytes as biocontrol agents of pre- and post-harvest diseases: Fundamentals, methods of application and future perspectives. *Microbiol. Res.* **2021**, *242*, 126612, doi:10.1016/j.micres.2020.126612.

[27] Strobel, G. The emergence of endophytic microbes and their biological promise. *J. Fungi* **2018**, *4*, doi:10.3390/jof4020057.

[28] Bolívar-Anillo, H.; Orozco-Sanchez, C.; da Silva Lima, G.; Franco dos Santos, G. Endophytic Microorganisms Isolated of Plants Grown in Colombia: A Short Review. *J. Microb. Biochem. Technol.* **2016**, *08*, 509-513, doi:10.4172/1948-5948.1000335.

[29] Redman Regina, S.; Dunigan David, D.; Rodriguez Rusty, J. Fungal symbiosis from mutualism to parasitism: who controls the outcome, host or invader? *New Phytol.* **2001**, *151*, 705-716.

[30] Orozco-Mosqueda, M. del C.; Santoyo, G. Plant-microbial endophytes interactions: Scrutinizing their beneficial mechanisms from genomic

explorations. *Curr. Plant Biol.* **2021**, *25*, 100189, doi:10.1016/j.cpb.2020.100189.

[31] Latz, M.A.C.; Jensen, B.; Collinge, D.B.; Jørgensen, H.J.L. Endophytic fungi as biocontrol agents: elucidating mechanisms in disease suppression. *Plant Ecol. Divers.* **2018**, *11*, 555-567, doi: 10.1080/17550874.2018.1534146.

[32] Hardoim, P.R.; van Overbeek, L.S.; Berg, G.; Pirttilä, A.M.; Compant, S.; Campisano, A.; Döring, M.; Sessitsch, A. The Hidden World within Plants: Ecological and Evolutionary Considerations for Defining Functioning of Microbial Endophytes. *Microbiol. Mol. Biol. Rev.* **2015**, *79*, 293-320, doi:10.1128/mmbr.00050-14.

[33] Santoyo, G.; Moreno-Hagelsieb, G.; del Carmen Orozco-Mosqueda, M.; Glick, B.R. Plant growth-promoting bacterial endophytes. *Microbiol. Res.* **2016**, *183*, 92-99, doi:10.1016/j.micres.2015.11.008.

[34] Hong, C.E.; Park, J.M. Endophytic bacteria as biocontrol agents against plant pathogens: current state-of-the-art. *Plant Biotechnol. Rep.* **2016**, *10*, 353-357, doi:10.1007/s11816-016-0423-6.

[35] Bolívar-Anillo, H.J.; González-Rodríguez, V.E.; Cantoral, J.M.; García-Sánchez, D.; Collado, I.G.; Garrido, C. Endophytic Bacteria *Bacillus subtilis*, Isolated from *Zea mays*, as Potential Biocontrol Agent against *Botrytis cinerea*. *Biology*. **2021**, *10(6)*, 492; https://doi.org/10.3390/biology10060492.

[36] Shelake, R.M.; Pramanik, D.; Kim, J.Y. Exploration of plant-microbe interactions for sustainable agriculture in CRISPR era. *Microorganisms* **2019**, *7*, 1-32, doi:10.3390/microorganisms7080269.

[37] Afzal, I.; Shinwari, Z.K.; Sikandar, S.; Shahzad, S. Plant beneficial endophytic bacteria: Mechanisms, diversity, host range and genetic determinants. *Microbiol. Res.* **2019**, *221*, 36-49, doi:10.1016/j.micres.2019.02.001.

[38] El-Sayed, A.S.A.; Akbar, A.; Iqrar, I.; Ali, R.; Norman, D.; Brennan, M.; Ali, G.S. A glucanolytic *Pseudomonas* sp. associated with *Smilax bona-nox* L. displays strong activity against *Phytophthora parasitica*. *Microbiol. Res.* **2018**, *207*, 140-152, doi:10.1016/j.micres.2017.11.018.

[39] Wang, Y.; Zeng, Q. gui; Zhang, Z. bin; Yan, R. ming; Zhu, D. Antagonistic bioactivity of an endophytic bacterium H-6. *African J. Biotechnol.* **2010**, *9*, 6140-6145, doi:10.5897/AJB10.258.

[40] Zhao, L.F.; Xu, Y.J.; Lai, X.H. Antagonistic endophytic bacteria associated with nodules of soybean (*Glycine max* L.) and plant growth-promoting properties. *Brazilian J. Microbiol.* **2018**, *49*, 269-278, doi:10.1016/j.bjm.2017.06.007.

[41] Melnick, R.L.; Zidack, N.K.; Bailey, B.A.; Maximova, S.N.; Guiltinan, M.; Backman, P.A. Bacterial endophytes: *Bacillus* spp. from annual crops as potential biological control agents of black pod rot of cacao. *Biol. Control* **2008**, *46*, 46-56, doi:10.1016/j.biocontrol.2008.01.022.

[42] Khalaf, E.M.; Raizada, M.N. Bacterial seed endophytes of domesticated cucurbits antagonize fungal and oomycete pathogens including powdery mildew. *Front. Microbiol.* **2018**, *9*, 1-18, doi:10.3389/fmicb.2018.00042.

[43] Misk, A.; Franco, C. Biocontrol of chickpea root rot using endophytic actinobacteria. *BioControl* **2011**, *56*, 811-822, doi:10.1007/s10526-011-9352-z.

[44] Bhusal, B.; Mmbaga, M.T. Biological control of Phytophthora blight and growth promotion in sweet pepper by *Bacillus* species. *Biol.*

Control **2020**, *150*, 104373, doi:10.1016/j.biocontrol.2020.104373.

[45] Agisha, V.N.; Kumar, A.; Eapen, S.J.; Sheoran, N.; Suseelabhai, R. Broad-spectrum antimicrobial activity of volatile organic compounds from endophytic *Pseudomonas putida* BP25 against diverse plant pathogens. *Biocontrol Sci. Technol.* **2019**, *29*, 1069-1089, doi:10.1080/09583157.2019.1657067.

[46] Kollakkodan, N.; Anith, K.N.; Nysanth, N.S. Endophytic bacteria from *Piper colubrinum* suppress *Phytophthora capsici* infection in black pepper (*Piper nigrum* L.) and improve plant growth in the nursery. *Arch. Phytopathol. Plant Prot.* **2020**, *0*, 1-23, doi:10.1080/03235408.2020.1818493.

[47] Iqrar, I.; Shinwari, Z.K.; El-Sayed, A.S.A.F.; Ali, G.S. Exploration of microbiome of medicinally important plants as biocontrol agents against *Phytophthora parasitica*. *Arch. Microbiol.* **2021**, doi:10.1007/s00203-021-02237-2.

[48] Munjal, V.; Nadakkakath, A.V.; Sheoran, N.; Kundu, A.; Venugopal, V.; Subaharan, K.; Rajamma, S.; Eapen, S.J.; Kumar, A. Genotyping and identification of broad spectrum antimicrobial volatiles in black pepper root endophytic biocontrol agent, *Bacillus megaterium* BP17. *Biol. Control* **2016**, *92*, 66-76, doi:10.1016/j.biocontrol.2015.09.005.

[49] Alsultan, W.; Vadamalai, G.; Khairulmazmi, A.; Saud, H.M.; Al-Sadi, A.M.; Rashed, O.; Jaaffar, A.K.M.; Nasehi, A. Isolation, identification and characterization of endophytic bacteria antagonistic to *Phytophthora palmivora* causing black pod of cocoa in Malaysia. *Eur. J. Plant Pathol.* **2019**, *155*, 1077-1091, doi:10.1007/s10658-019-01834-8.

[50] Cheffi, M.; Bouket, A.C.; Alenezi, F.N.; Luptakova, L.; Belka, M.; Vallat, A.; Rateb, M.E.; Tounsi, S.; Triki, M.A.; Belbahri, L. Olea europaea l. Root endophyte bacillus velezensis oee1 counteracts oomycete and fungal harmful pathogens and harbours a large repertoire of secreted and volatile metabolites and beneficial functional genes. *Microorganisms* **2019**, *7*, 1-29, doi:10.3390/microorganisms7090314.

[51] Abraham, A.; Philip, S.; Jacob, M.K.; Narayanan, S.P.; Jacob, C.K.; Kochupurackal, J. Phenazine-1-carboxylic acid mediated anti-oomycete activity of the endophytic *Alcaligenes* sp. EIL-2 against *Phytophthora meadii*. *Microbiol. Res.* **2015**, *170*, 229-234, doi:10.1016/j.micres.2014.06.002.

[52] Ngo, V.A.; Wang, S.L.; Nguyen, V.B.; Doan, C.T.; Tran, T.N.; Tran, D.M.; Tran, T.D.; Nguyen, A.D. *Phytophthora* antagonism of endophytic bacteria isolated from roots of black pepper (Piper nigrum L.). *Agronomy* **2020**, *10*, 1-15, doi:10.3390/agronomy10020286.

[53] Segaran, G.; Sathiavelu, M. Fungal endophytes: A potent biocontrol agent and a bioactive metabolites reservoir. *Biocatal. Agric. Biotechnol.* **2019**, *21*, 101284, doi:10.1016/j.bcab.2019.101284.

[54] Fouda, A.H.; Hassan, S.E.D.; Eid, A.M.; Ewais, E.E.D. Biotechnological applications of fungal endophytes associated with medicinal plant *Asclepias sinaica* (Bioss.). *Ann. Agric. Sci.* **2015**, *60*, 95-104, doi:10.1016/j.aoas.2015.04.001.

[55] Hanada, R.E.; Pomella, A.W. V.; Costa, H.S.; Bezerra, J.L.; Loguercio, L.L.; Pereira, J.O. Endophytic fungal diversity in *Theobroma cacao* (cacao) and *T. grandiflorum* (cupuaçu) trees and their potential for growth promotion and biocontrol of black-pod disease. *Fungal Biol.* **2010**, *114*, 901-910, doi:10.1016/j.funbio.2010.08.006.

[56] Mitchell, A.M.; Strobel, G.A.; Moore, E.; Robison, R.; Sears, J. Volatile antimicrobials from *Muscodor crispans*, a

novel endophytic fungus. *Microbiology* **2010**, *156*, 270-277, doi:10.1099/mic.0.032540-0.

[57] Mathew, S.K.; Mary, C.F.G.; Gopal, K.S.; Girija, D. Antagonistic Activity of Endophytic *Trichoderma* against Phytophthora Rot of Black Pepper (*Piper nigrum* L.). *J. Biol. Control* **2011**, *25*, 48-50, doi:10.18311/jbc/2011/3840.

[58] Bae, H.; Roberts, D.P.; Lim, H.S.; Strem, M.D.; Park, S.C.; Ryu, C.M.; Melnick, R.L.; Bailey, B.A. Endophytic *Trichoderma* isolates from tropical environments delay disease onset and induce resistance against *Phytophthora capsici* in hot pepper using multiple mechanisms. *Mol. Plant-Microbe Interact.* **2011**, *24*, 336-351, doi:10.1094/MPMI-09-10-0221.

[59] Miles, L.A.; Lopera, C.A.; González, S.; de García, M.C.C.; Franco, A.E.; Restrepo, S. Exploring the biocontrol potential of fungal endophytes from an Andean Colombian Paramo ecosystem. *BioControl* **2012**, *57*, 697-710, doi:10.1007/s10526-012-9442-6.

[60] Tellenbach, C.; Sumarah, M.W.; Grünig, C.R.; Miller, J.D. Inhibition of *Phytophthora* species by secondary metabolites produced by the dark septate endophyte *Phialocephala europaea*. *Fungal Ecol.* **2013**, *6*, 12-18, doi:10.1016/j.funeco.2012.10.003.

[61] Park, Y.H.; Chung, J.Y.; Ahn, D.J.; Kwon, T.R.; Lee, S.K.; Bae, I.; Yun, H.K.; Bae, H. Screening and characterization of endophytic fungi of *Panax ginseng* Meyer for biocontrol activity against ginseng pathogens. *Biol. Control* **2015**, *91*, 71-81, doi:10.1016/j.biocontrol.2015.07.012.

[62] Terhonen, E.; Sipari, N.; Asiegbu, F.O. Inhibition of phytopathogens by fungal root endophytes of Norway spruce. *Biol. Control* **2016**, *99*, 53-63, doi:10.1016/j.biocontrol.2016.04.006.

[63] Sreeja, K.; Anandaraj, M.; Bhai, R.S. In vitro evaluation of fungal endophytes of black pepper against *Phytophthora capsici* and *Radopholus similis*. *J. Spices Aromat. Crop.* **2016**, *25*, 113-122.

[64] Wang, G.; Liu, Z.; Lin, R.; Li, E.; Mao, Z.; Ling, J.; Yang, Y.; Yin, W.B.; Xie, B. Biosynthesis of Antibiotic Leucinostatins in Bio-control Fungus *Purpureocillium lilacinum* and Their Inhibition on *Phytophthora* Revealed by Genome Mining. *PLoS Pathog.* **2016**, *12*, 1-30, doi:10.1371/journal.ppat.1005685.

[65] Sánchez-Ortiz, B.L.; Sánchez-Fernández, R.E.; Duarte, G.; Lappe-Oliveras, P.; Macías-Rubalcava, M.L. Antifungal, anti-oomycete and phytotoxic effects of volatile organic compounds from the endophytic fungus *Xylaria* sp. strain PB3f3 isolated from *Haematoxylon brasiletto*. *J. Appl. Microbiol.* **2016**, *120*, 1313-1325, doi:10.1111/jam.13101.

[66] Sánchez-Fernández, R.E.; Sánchez-Fuentes, R.; Rangel-Sánchez, H.; Hernández-Ortega, S.; López-Cortés, J.G.; Macías-Rubalcava, M.L. Antifungal and antioomycete activities and modes of action of isobenzofuranones isolated from the endophytic fungus *Hypoxylon anthochroum* strain Gseg1. *Pestic. Biochem. Physiol.* **2020**, *169*, 104670, doi:10.1016/j.pestbp.2020.104670.

Section 3

Prevalence and Recognition of *Phytophthora*

Chapter 5

Phytophthora Diseases Prevalence, Its Effects and Controls in Ghana

Benedicta Nsiah Frimpong, Samuel Oteng Ampadu, Allen Oppong, Isaac Nunoo and Lydia Brobbey

Abstract

The success of the UN Sustainable Development Goals in reducing hunger and poverty is limited by crop losses. Globally, plant pests and diseases account for 40% yield losses which threatens food and nutrition security, livelihoods of citizenry and erode the resources of local and national economies. *Phytophthora* diseases are among the most important diseases in sub-Saharan Africa which result in severe socio-economic consequences. Roots and tubers and cash commodity crops are important staples and foreign exchange earner crops in Ghana which are significantly challenged by the incidence and severity of *Phytophthora* diseases. To ensure food availability, safeguard the local financial ecosystem and protect the environment, innovative and sound management practices are needed and this chapter reviews the different *Phytophthora* diseases on crops; more specifically with (cocoa and taro as case studies), the consequences and available management options that can be applied to manage the disease situation in Ghana.

Keywords: economic loss, incidence, severity, Cocoa, Taro, sustainable development, Ghana

1. Introduction

1.1 Origin of *Phytophthora* diseases

Phytophthora is a genus of filamentous Oomycetes, within the Kingdom Chromista which is also referred to as Kingdom Stramenopila [1–3]. There are several species within the class Oomycetes of which over 120 species are known [4] and could either be soil or water borne. Morphologically, they bear the resemblance of fungi with both sexual and asexual spores. Most are pathogens causing disease in a large range of plant hosts. The pathogens not only cause economic damage to crops but to the natural ecosystem as well. They affect both traditional and non-traditional agricultural crops, floricultural plants such as ornamentals and forest plants and they are pervasive in soil and water globally [3]. Due to their economic and environmental impact, there is expanding interest in *Phytophthora* genetics and genomics, resulting in the recent releases of genome sequences of *P. ramorum*, *P. sojae, P. infestans, P. capsici and P. litchi* [5–8].

The identification of gene families encoding classes of toxins, elicitors, and effectors shared among the *Phytophthora* species is critical to understanding the disease process. The most devastating specie worldwide is the *P. infestans* which in history caused huge damage to Irish Potatoes in 1845 as a result of Potato Late Blight outbreak thus, causing great famine in the Irish land with about 25% of its population starving and evacuating [3]. *P. infestans* are noted to affect the Solanaceous plants while the rest may either be host specific or attack varied host plants. Below are some selected species for the temperate and tropical regions of the globe; their target host plants and signs and symptoms (**Table 1**).

1.2 *Phytophthora* diseases prevalence: Evidence from Ghana

1.2.1 Phytophthora *disease in cocoa production*

In Ghana, the earliest form of *Phytophthora* disease was caused by the pathogen P. megakarya. P. megakarya is native and pervasive in the Western and Central parts of Africa. Known to have spread from the West form Cameroon to Ghana and Cote D'Ivoire through Nigeria and Togo and Southwards to Gabon and Equatorial Guinea [13]. *Phytophthora* megakarya is the most destructive fungal pathogen on cocoa production in Ghana [13]. The disease has been in Ghana for many years but on other alternative hosts [16]. It was originally identified in Nigeria in 1979 [17], reported in Togo in 1982 [18], and was subsequently reported in Ghana in 1985 [16]. Though the incidence of *Phytophthora* disease was originally reported in Ghana by 1985, Darkwa (1981) concluded that P.megakarya probably occurred before 1980 until it was officially reported in 1985 at Akomadan-Ashanti Region.

According to Tsopmbeng et al. [8] an isolate of *Phytophthora* from *Mimusops elengi* at Aburi in Ghana was distinctly different from what was hitherto referred to as cocoa or G-isolate. Turner et al. [8] further reported that the Mimusops isolate produced oospores in mixed culture with the cocoa isolate. Presently, the G-isolate has now been identified as P. palmivora and the N-isolate as P. megakarya [16, 19]. Until 1985, *Phytophthora* palmivora was the only known causal agent for *Phytophthora* pod rot (black pod) disease in Ghana. The appearance of *Phytophthora* megakarya in 1985 in Ghana added a new dimension to the disease complex of cocoa in the country. Similarly, studies on black pod diseases by [20] confirmed that some parts of Volta Region of Ghana consistently had the predominant type caused by P. megakarya. This is plausible due to the fact that the region shares boundary with Togo, a country predominantly affected by P. megakarya species [16].

1.2.2 Phytophthora *disease in Taro production*

It is also important to highlight that not only has the prevalence of *Phytophthora* affected cocoa production after its earliest occurrence but also the production of Taro. The production of taro in Ghana, in recent times, has been affected by the taro leaf blight caused by *Phytophthora* colocasiae which has also been reported to have threatened the sustainability of taro production globally [21, 22]. In Ghana, [23] reported the presence of the disease after similar reports in Nigeria and Cameroon. The disease affects all parts of the crop including the leaves, corms, petioles and cormels, resulting in extensive damage of the foliage and reduced yield [24]. It has therefore become a limiting factor to taro production in all taro growing countries.

Taro (*Colocasia esculenta* var., antiquorum) is one of the most important food crops in Ghana [25]. It is a hunger crop and cultivated in almost all the ten regions of the country [26]. The corm is used for flour for bakery and in the preparation of local dishes. The corm is also high in carbohydrates [27]. The leaves can be eaten as vegetable in the country, and it is an excellent source of vitamins. A lot of village folks depend on

No.	*Phytophthora* species	Target host plant	Signs and symptoms	Relevant literature
1.	*P. alni*	Alder	Rot of the root and collar	[3]
2.	*P. cactorum*	Rhododendrons, Azaleas, hardwood, apple, pear, strawberry.	Causes root rot in rhododendrons, azaleas and other related species. It affects woody trees causing cancer and other economic important fruits.	[3]
3.	*P. cinnamomi*	Woody ornamental: arborvitae, azalea, Chamaecyparis; dogwood Taxus, white pine, American chestnut and Eucalyptus (jarrah).	Root rot and seedling mortality	[3]
4.	*P. cambivora*	Chestnut, apple, pear, peach, almond	Root rot to forest trees, crown dieback, flame blot at the collar region, discoloration and lesions at the growing point and wilting.	[3, 9]
5.	*P. citrophthora*	citrus, pistachio, peach and ornamental species	Root rot, it's a soil-borne disease of citrus causing gummosis and brown rot with pungent smell.	[3, 9, 10]
6.	*P. cryptogea*	Ornamental species	Causes collar rot with reduction in foliage. First reported to cause gummosis in citrus in Tunisia.	[3, 11]
7.	*P. kernoviae*	Beech, rhododendron, trees and shrub species	Restricted to the UK and Ireland environment. Causes abnormal leaf fall, necrosis and whole plant dieback.	[3, 12]
8.	*P. megakarya*	Cocoa	Virulent specie which causes pod rot resulting in black pod diseases. It reduces yield and accounts for greater losses of cocoa especially in West and Central Africa. About 60–100% losses incurred if not managed.	[3, 13, 14]
9.	*P. nicotianae*	Tobacco, onions, cotton, ornamental species, coconut and pineapple	Cause diseases	[3]
10.	*P. palmivora*	in coconuts and betel nuts, palm species, papaya, cocoa	Fruit rot, stem and root rots in other tropical fruits. Causes pod rot in cocoa in most growing countries. It causes about 20–30% yield losses	[3, 14]
11.	*P. infestans*	Potato, tomato	Infects the above ground part of the plant and occurs at any stage of the plant development. Elongated, dark lesions on tomato branches.	[15]
12.	*P. citricola*	Avocado, conifers	Causes crown and canker rots	[3]

Table 1.
Some selected species and the affected host plants.

this crop for their livelihood. Farmers obtain regular income from the production as well as food for the family [26]. Despite, the importance of taro as an important food security crop in Ghana; its production is hampered by a leaf blight disease caused by *Phytophthora* colocasiae. Marian Raciborski first described *Phytophthora* colocasiae in 1900 from Java. It was first reported in Ghana in 2012 [28]. It is the most destructive fungal disease responsible for heavy yield losses (25 to 50%) of taro [27]. In addition, this pathogen causes a serious postharvest decay of taro corms.

1.2.3 Method of isolating and identification of Phytophthora spp

Proper plant diseases identification is critical and it forms the basis for population genetics, epidemiological studies and development of effective control mechanisms. This review reports on how authors have isolated *phytophthora spp* in cocoa and taro respectively.

1.2.3.1 P. Megakarya *isolation*

Detailed account on experiment conducted by [29] investigating shade trees as alternative host of P. megakarya is given as follows. The team conducted an experiment on a 5-hectare cocoa field at erstwhile Brong Ahafo region, precisely Bechem which was planted in 1984 with two hybrids; T79/501 x Amel, and T60 x Na45 respectively. Cocoa plants in the test field were largely infected with P. megakarya. Forty-eight out of the fifty isolates recovered from the cocoa pods representing ninety-six percent were found to be *P. megakarya* with only two identified as *P. palmivora*.

The team identified 34 shade trees at the test site so roots with no visible lesions of approximately 1-2 cm thick were collected from a depth ranging from 20 to 50 cm. separate samples were placed in black polythene bags and refrigerated at 4oC for up to 2 weeks before isolations were done. Samples were taken in two month intervals in the 1996/97, 1997/98 and 1998/99 cropping calendar (June, August, October, December, February and April). During the isolation process, the bulk soil and pieces of other root parts were thoroughly washed with running water. A razor blade was used to cut about 1–2 were of the roots and washed in three separate sterile distilled water. The surface sterilized immersing for 5 min in a 10% sodium hypochlorite solution and wiped dry on a paper towel. The roots were again washed for an hour in sterile distilled water on a flask shaker. In all 100 root pieces were cut from each test tree and sub divided into two groups. The isolation methods involved two techniques of "baiting with cocoa pod husks and direct plating on P. megakarya and P. palmivora agar (PPMA)" [30]. Isolates were identified on the basis of three parameters; growth rates, colony morphology and sporangium features. Out of 34 shade trees tested, P. megakarya was recovered from four of the roots from the shade trees after three consecutive years. P. megakarya was isolated most frequently in the wet season than the dry.

1.2.3.2 P. colocasiae *isolation*

A survey was conducted by [31] during the 2019 rainy (July to November, 2019) and dry seasons (November, 2019 to February, 2020) in Sunyani and Dorma-Central Municipalities to assess the incidence and severity of taro blight disease in these zones. The team collected randomly sampled from infected leaves and petioles showing sign like "the development, exudation and oozing of amber, reddish-brown or bright-orange droplets from both sides of the leaf margins, water-soaked necrotic areas, which have combined into large lesions with white powdery appearance and blighted leaf blade. The sample was take to the University of Energy and Natural Resources Lab in Ghana for isolation and purification of the pathogen. The isolation was carried out under a Laminar flow hood. The diseased part of the taro leaves and petiole was cut. The pieces were surface sterilized in 70% ethanol for a minute and carefully washed in three exchanges of distilled water. The pieces were blotted dry on Whatman paper for 2 minutes and plated on potato dextrose agar (PDA, Oxoid, England) at the 28°C for seven days. They were examined daily for the development of mycelial growth. The isolation process was reproduced three times. The mixed population cultures were sub- cultured by transferring hyphal tip from the mycelium edge onto a new prepared PDA medium using flamed inoculation needle to purify it.

Wet mount from the pure cultures was prepared to identify the pathogen morphologically. By using bi-nuclear microscope, the characters of the putative pathogen such as hyphae type, shape of sporangia, micro and macro conidia were also examined morphologically and the characteristics compared to a standard established identification protocols by [32, 33].

1.3 Symptoms and distribution of virulent *phytophthora* diseases in Ghana

1.3.1 Phytophthora *disease cycle and environmental parameters for disease incidence*

Direct correlation has been established between black pod disease incidence and weather condition. Thus black pod disease has been seen to be highly influenced by environmental factors and several studies [13, 34] have confirmed the role played by climate variability in the prevalence of black pod disease caused by *phytophthora* species. Akrofi et al. [35] reported that the disease develops well under frequent precipitation, high relative humidity and low temperature. Under high and regular precipitation regimes, P. megakarya is reported to result in a total yield loss in Cameroon where no action was taken [13, 35]. Under similar conditions; Asare-Nyako and Dakwa [34] reported losses in the range of 60 to 100% in Ghana. Asare-Nyako and Dakwa [34] emphasized that, the black pod disease in Ghana developed quickly during the day when the relative humidity stayed above 80% under shady cocoa and the frequency and amount of rainfall influenced the intensity of the disease development. Asare-Nyako and Dakwa [34] reiterated that the peak level of infection varied yearly between location and with the rainfall pattern.

In Ghana, black pod disease caused by P. megakarya is usually severe between August and October [16, 36]. The topmost phases of disease occurrence provide rich information in predicting disease development trends and serve as an important disease management tool. The developmental stages during *phytophthora* disease cycle in cocoa is presented in **Figure 1**.

P. colocasiae survives under high temperatures and humidity, in wet areas and plots that are densely planted [12]. Study by [25] in Aowin Suaman district in the Western Region of Ghana; a tropical rainforest with monthly temperature of 27°C and annual rainfall between 1500 and 1800 millimeters, recorded high incidence of Taro leaf blight; 99% as the described condition favored the spread of the disease.

1.3.2 Symptoms and distribution of P. megakarya

Phytophthora disease incidence and crop losses vary from one locality and farm to another [35] and fluctuate across seasons [20]. P. megakarya infects every

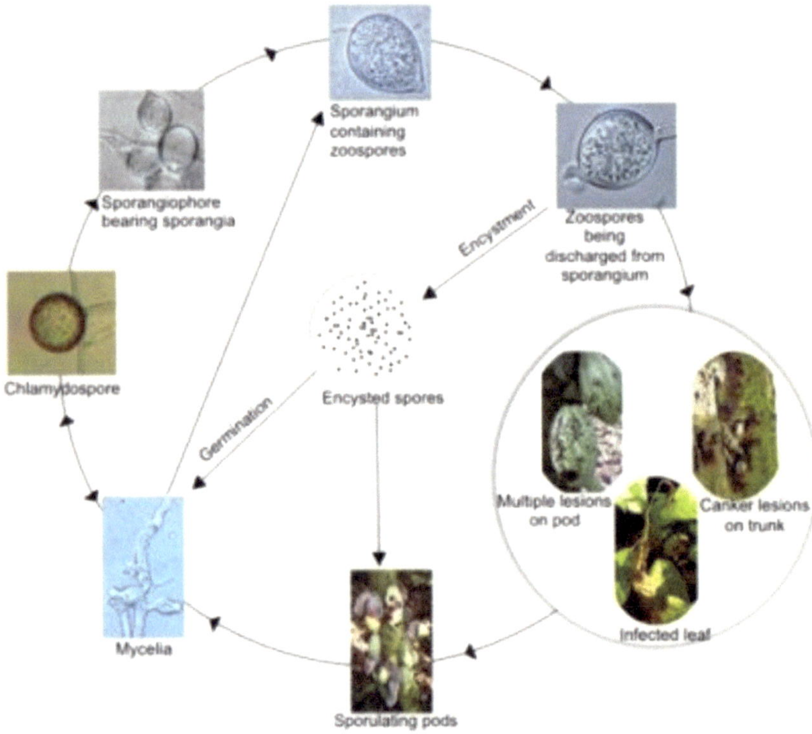

Figure 1.
Phytophthora *disease cycle in cocoa. Photo credit: Akrofi et al. [13].*

developmental stage and every part of the cacao plant under wet and humid conditions. Infection of seedlings leads to leaf blight and root rot in nurseries, while infections of the stem, chupons and branches lead to cankers. While every stage of pod development is susceptible to infection, immature pods are the most susceptible. Pod infection also leads to pod rot [13]. In Ghana, P. megakarya form stem cankers very rapidly. Unlike P. palmivora cankers which are usually distributed normally on the tree trunk, P. megakarya cankers tend to be concentrated on the lower parts of the stem close to the ground though it affects all parts of the tree [36]. Due to this, treating with chemicals become difficult and unproductive.

In Ghana, [13] found that P. megakarya has spread from Akomadan and Bechem where it was first reported in 1985 into 50 more administrative districts in the six cocoa growing regions of Ghana covering an approximate area of 75,298 km2. They further noted that the current distribution in the country is as follows: Ashanti, 13 districts (17,676 km2); Brong Ahafo Region, 10 districts (10,422 km2); Central Region, 4 districts (5900 km2); Eastern, 7 districts (7760 km2); Western, 12 districts (25,698 km2) and Volta, 6 districts 7843 km2. The corresponding percentage areas infested in the regions are 23.5%, 13.8%, 7.8%, 10.3%, 34.1% and 10.4% respectively [13]. Pictures of infected cocoa pod showing symptoms is presented in **Figure 2**.

1.3.3 Symptoms and spread of Phytophthora *colocasiae*

Taro (*Colocasia esculenta* (L.) Schott) suffers attacks from several pathogens, among which *Phytophthora* colocasiae, Racib, associated with the Taro leaf blight being the most destructive. The disease is associated with 90% and 50% loss in leaf

Figure 2.
Symptoms of P. megakarya in cocoa in Ghana. Photo credit: Akrofi et al. [13]. (a) multiple lesions on cocoa pod; (b) coalescing lesions; (c) abundant sporangia indicated by the arrow; (d) diverse infection phases on the cocoa; (e) distal infection; (f) proximal infection; (g) lateral infection; (h) canker lesions prior to scraping and (i) canker lesions once scraped displaying scarlet colouration.

and corm yield of taro, respectively [22]. *Phytophthora* colocasiae is disseminated by infected vegetative plant parts and possibly contaminated soil [21]. Conventionally, variations in *Phytophthora* species have been detected on host differential, biochemical test, morphological and molecular level characterizations [37]. According to studies, the foliar pathogen has spread across Africa, East Asia, the Americas, the Caribbean, and the Pacific, as well as all other taro-growing regions of the world, with varying degrees of severity [28, 38].

Figure 3.
A. Asymptomatic leaf; B. Symptomatic leaf. Photo credit: Abdulai et al. [31].

Figure 4.
A. Water-soaked appearance on plant; B. Light exudate from both sides of the water-soaked leaf margins. Photo credit: Abdulai et al. [31].

In terms of its dissemination, [31] indicated that windy rains and splashing water from irrigation or running water are two ways that the sporangia on the infected plant surface are quickly disseminated. The pathogen can grow in the soil as an encysted zoospore with thick cover layers or as chlamydospores in the absence of the host for several months as a survival mechanism under dry stress conditions [28, 39]. Blight of the leaf blade is the most visible symptom of the disease; other symptoms include postharvest rot of the corm and rotting of the petiole in susceptible varieties [40]. Early plant leaf infection is most common in areas where there is sufficient accumulation of guttation droplets, dew, or rainfall. The pathogen sporangia usually appear on infected leaves as small, brown, water-soaked necrotic areas that quickly coalesce into large lesions from which yellow exudates emerge, followed by defoliation and plant death within a few weeks after infection [25, 28].

The fluctuating day/night cycle influences the development of specific symptoms. Cool night temperatures encourage lesion expansion with 3–5 mm wide water-soaked margins that dry out during the day and return to water-soaked status at night, resulting in zonation around the necrotic lesion that is easily visible when viewed from the bottom of the infected leaf [41]. Some symptomatic and asymptomatic plant parts are shown in **Figures 3–5**.

Figure 5.
A–D. Fresh and dried Amber, bright-Orange or reddish-Brown exudate on Taro leaves. Photo credit: Abdulai et al. [31].

2. Social and economic impact of infectious *Phytophthora* diseases in Ghana

2.1 Impact of *Phytophthora* megakarya

Though cocoa is native of South America, the bulk of the beans production comes from Africa with Ghana being the second largest world's producer after Cote D'Ivoire [42, 43] With Ghana's position in the International cocoa production and export markets [42], cocoa contribution to the nation's economic growth is limited by high yield losses resulting from *Phytophthora* disease infections [13]. *Phytophthora* palmivora which accounted for pod losses of less than 30% was the only known causal agent of black pod disease of cocoa in Ghana prior to 1985 A. P. megakarya causes yield losses as high as 60–100% in Ghana according to a report by [44]. P. megakarya has become the main yield-limiting factor for cocoa production in affected areas [36], rapidly surpassing the importance of P. palmivora. The emergence of P. megakarya has had dramatic social and economic consequences in cocoa producing countries in West and Central Africa including Ghana, clearly demonstrating the scale of damage that it may cause in case it spreads into other cocoa producing territories.

Particularly in Ghana, it was reported that some cocoa farms were neglected or abandoned and, some cocoa farmers switched over to cultivate vegetables and other crops because of P. megakarya infections on their cocoa farms [35, 36].

A report by COCOBOD in 2014 is also indicative that Ghana lost over 25% (212,500 MT) of its annual output of 850,000 MT of cocoa beans to black pod disease, representing a revenue loss of about GH¢7.5 million in 2012. P. megakarya still remains an invasive species in Ghana and was reported to be spreading in the Ghanaian cocoa belt towards the border with Cote d'Ivoire [29]. A study by [16] noted that several national programmes, including the National Cocoa Pests and Diseases Control Programme (CODAPEC), were instituted by the Ghanaian government in which P. megakarya infected farms were sprayed with fungicides at the expense of the government. The money spent on these programs could have been better spent on improving the lives of farmers. In addition, [36] noted that in view of the severity of P. megakarya mediated black pod during the disease-conducive period (July–October), some famers in Ghana attached some belief to its incidence due to the devastating nature; thinking that it was a strange disease caused by evil forces or the effects of the Volta Lake [35] which has influenced farmers to adopt wrong attitudes towards its control.

2.2 Impact of *Phytophthora* colocasiae

Globally, it is generally believed that diseases decrease agricultural productivity by more than 10%, which is comparable to half a billion tonnes of total food produced each year [40]. The impact of fungal diseases on crop production has been well explained by [31] who reported that, when fungal diseases are properly controlled on five (5) major crops alone, more than 600 million people could be fed each year in the world. Taro plant is not an exception, as it is known to be infected by more than ten serious pests and diseases caused by a number of insect pests and pathogens across the globe [45]. Among all the disease-causing agents in taro plants, P. colocasiae, which causes leaf blight of taro is known to be the most important. This pathogen has been reported widely for causing leaf yield loss of 95% and 50% in postharvest rot of corm yield and quality [39, 46].

It is believed that P. colocasiae is disseminated by means of vegetative propagation materials [28] and the case may not be different in Ghana. There are no accredited supply centers for planting materials in the country, and farmers rely on families, neighbors and open market for their supplies of planting materials, which may be coming from already infested fields. The constraints of taro blight disease to productivity of taro have been acknowledged in the West-African Sub Region [28, 38, 47]. The disease poses serious threats to global food security as well as economic hardship to the people in these taro producing regions of the world. In Ghana, apart from the three northern regions, taro production is mainly carried out in the southern part of the country. A few research works have been reported so far on Taro [47–50]. Even then, the focus had been on the profitability of the taro enterprise. More studies such as ours reporting on the incidence and severity of P. colocasiae are needed to provide valuable data to inform interventions towards the management of taro blight in the country.

3. Management of *phytophthora* diseases in Ghana

3.1 Management of *Phytophthora* megakarya

Huge losses resulting from P. megakarya and associated management cost pose serious threats on the socio-economic development of cocoa growing countries in terms of their financial resources such as Ghana. Timely, more integrated and sustainable practices which involve the use of resistant varieties, chemical application, quarantining germplasm received outside the country, cultural and biological control are imperative [13, 35]. To effectively prevent disease caused by P. megakarya, these integrated control strategies must be employed on time. Planting material movement from one location to another within Ghana account for the quick spread of the pod rot pathogens. The amalgamation of cultural and chemical methods in Ghana has proven to be effective against P. megakarya. Cultural practices present a cost effective way of managing plant diseases as it provides the right ecosystem for effective performance of fungicides. Cultural practices alone, including judicious shade management, pruning, removal of basal chupons, mistletoes and frequent harvesting, can be sufficient to control P. palmivora [34, 36]. Cultural practices are not only essential for increasing yield, but also provide the right environment for the efficient performance of recommended fungicides [36]. Frequent harvesting, for instance, saves partly infected mature pods and reduces sources of sporangial inoculum while shade management; opening up the canopy and reducing basal chupons, enhances air circulation in the cocoa farm, thereby reducing disease incidence [51]. Iwaro et al. [51] noted that at least six applications are required in one black pod sea.

The recommendation of 3-weekly fungicide spraying in Ghana (son (May–October). This rather high frequency of spraying, coupled with the ever-increasing cost of inputs (labour and fungicides) and the lack of knowledge in techniques for effective spraying, make the adoption of chemical control very low. Four-weekly spraying of either metalaxyl and copper-1-oxide (Ridomil 72 plus) or cuprous oxide (Nordox 75) combined with cultural practices had been found effective against *Phytophthora* megakarya in researcher-managed trials [52]. This spraying regime reduces the number of sprays per season to five.

3.2 Management of *Phytophthora* colocasiae

To be able to control or manage taro blight disease, which usually limits the productivity of this crop, it is important that the pathogen is isolated from the diseased tissues and characterized. On that basis, the pathogen was successfully isolated and identified morphologically as P. colocasiae based on the important characters of the pathogen using standard Mycological identification keys according to [32, 33]. The sporangia are ovoid to ellipsoid with a well-defined narrow semi-papillate structure and are usually formed at the end of unbranched or casually branched sporangiophores at the edge of necrotic lesions. The sporangium is normally segregated from sporangiophores by the rain, leaving a small pedicel that is attached to their base [53], signifying the important role rain plays in the pathogen dispersal.

The incidence and severity of the disease are closely linked to the ability of the pathogen to be dispersed from one place to the other and hence the reason for the varied incidence and severity of the disease across the various fields in the communities'/farmers' fields. Management practices of taro leaf blight include hygienic practices, use of disease-free planting materials, wide spacing between plants when planting, clearing, removal and burning of infected debris (leaves) during the initial stage of disease development, separating the diseased plant from the healthy ones, planting near forest plantations which can serve as a barrier to disease transmission to the taro plants [54, 55].

Singh et al. [40] in his study was able to avoid serious taro blight disease in his field by planting during the dry season. Appropriate timing of planting is therefore recommended. Biological control methods such as the use of microorganisms, eg. *Pseudomonas fluorescens*, Trichoderma viride have also been applied [56]. Chemical control involving the use of systemic and protectant fungicides such as phosphorus acid (Foschek); copper (e.g. copper oxychloride); Mancozeb (e.g. Dithane M45) and metalaxyl (e.g. Ridomil Gold MZ) has successfully been used to control taro blight disease [46, 55]. Lastly, the use of the most effective and promising management strategy is the utilization of resistant taro cultivars [40] some of which were recently released by Scientists at CSIR-Crops Research Institute, Kumasi, Ghana.

4. Conclusion and recommendation

Despite the widespread distribution of *Phytophthora* diseases and resulting crop harm across the globe, there is a scarcity of knowledge about the diseases' incidence and intensity in Ghana. However, to efficiently establish a long-term management program for its control for farmers in Ghana to increase productivity, there is a need for more research to determine factors likely to limit the productivity of crops affected by *Phytophthora* disease. The continuous development of improved and high yielding varieties that are resistant or tolerant to Phythothora diseases should be intensified for all crops.

Conflict of interest

None.

Author details

Benedicta Nsiah Frimpong[1*], Samuel Oteng Ampadu[1], Allen Oppong[1], Isaac Nunoo[2,3] and Lydia Brobbey[1]

1 CSIR - Crops Research Institute, Kumasi, Ghana

2 Rural Education for Agricultural Development International, Kumasi, Ghana

3 Kwame Nkrumah University of Science and Technology, Kumasi, Ghana

*Address all correspondence to: benenash@yahoo.co.uk

References

[1] Bollmann, S. R., Fang, Y., Press, C. M., Tyler, B. M., and Grünwald, N. J. (2016). Diverse evolutionary trajectories for small RNA biogenesis genes in the oomycete genus Phytophthora. Frontiers in plant science, *7*, 284.

[2] Tyler, B. M., Tripathy, S., Zhang, X., Dehal, P., Jiang, R. H., Aerts, A. and Boore, J. L. (2006). Phytophthora genome sequences uncover evolutionary origins and mechanisms of pathogenesis. Science, *313*(5791), 1261-1266. https://doi.org/10.1126/science.1128796

[3] Giesbrecht, M. and Ridge, G. (2014). Phytophthora species (Root and crown rots of woodyplants). https://wiki.bugwood.org/Phytophthora_species_(Root_and_crown_rots_of_woody_plants) . Accessed: 2021/06/04

[4] Kroon, L. P., Brouwer, H., de Cock, A. W., and Govers, F. (2012). The genus Phytophthora anno 2012. Phytopathology, *102*(4), 348-364. https://doi.org/10.1094/phyto-01-11-0025

[5] Haas, B. J., Kamoun, S., Zody, M. C., Jiang, R. H., Handsaker, R. E., Cano, L. M., and Nusbaum, C. (2009). Genome sequence and analysis of the Irish potato famine pathogen Phytophthora infestans. Nature, *461*(7262), 393-398. https://doi.org/10.1038/nature08358

[6] Jing, M., Guo, B., Li, H., Yang, B., Wang, H., Kong, G. and Wang, Y. (2016). A *Phytophthora sojae* effector suppresses endoplasmic reticulum stress-mediated immunity by stabilizing plant Binding immunoglobulin Proteins. Nature Communications, *7*(1), 1-17.

[7] Lamour, K. H., Stam, R., Jupe, J., and Huitema, E. (2012). The oomycete broad-host-range pathogen Phytophthora capsici. Molecular plant pathology, *13*(4), 329-337. https://doi.org/10.1111/j.1364-3703.2011.00754.x

[8] Turner, P. D. (1960). Strains of Phytophthora palmivora (Butl.) Butl. From Theobroma cacao L.: isolates from West Africa. Transactions of the British Mycological Society, *43*(4), 665-IN7. https://doi.org/10.1016/S0007-1536(61)80035-1

[9] CABI. (2019). Invasive Species Compendium: *Phytophthora cambivora* (root rot of forest trees). https://www.cabi.org/isc/datasheet/40956 . Accessed: 2021/06/08

[10] Ippolito, A., Schena, L., Nigro, F., Soleti ligorio, V., and Yaseen, T. (2004). Real-time detection of Phytophthora nicotianae and P. citrophthorain citrus roots and soil. European Journal of Plant Pathology 110, 833-843 (2004). https://doi.org/10.1007/s10658-004-5571-9

[11] Boughalleb-M'hamdi, N., Benfradj, N., Migliorini, D., Luchi, N. and Santini, A. (2018). *Phytophthora nicotianae* and *P. cryptogea* causing gummosis of citrus crops in Tunisia. Trop. plant pathol. 43, 36-48 (2018). https://doi.org/10.1007/s40858-017-0180-2

[12] CABI. (2020). Invasive Species Compendium: *Phytophthora kernoviae. https://www.cabi.org/isc/datasheet/40972.* Accessed: 2021/06/08

[13] Akrofi, A. Y., Amoako-Atta, I., Assuah, M., and Asare, E. K. (2015). Black pod disease on cacao (Theobroma cacao, L) in Ghana: spread of Phytophthora megakarya and role of economic plants in the disease epidemiology. Crop Protection, *72*, 66-75. https://doi.org/10.1016/j.cropro.2015.01.015

[14] Shahbandeh, M. (2021). Production of cocoa beans in Ghana 2012/2013-2020/2021. https://www.statista.com/statistics/497844/production-of-cocoa-beans-in-ghana/, Accessed: 2021/06/10

[15] Nelson S. C. (2008). Late Blight of Tomato (*Phytophthora infestans*). Plant Disease. August, 2008. PD-45. https://www.ctahr.hawaii.edu/oc/freepubs/pdf/PD-45.pdf . Accessed: 2021/06/09

[16] Opoku, I. Y., Akrofi, A. Y., and Appiah, A. A. (2007). Assessment of sanitation and fungicide application directed at cocoa tree trunks for the control of Phytophthora black pod infections in pods growing in the canopy. European Journal of Plant Pathology, *117*(2), 167-175. https://doi.org/10.1007/s10658-006-9082-8

[17] Fagbohun, E. D., Falegan, C. R., and Arowolo, J. A. (2016). Phytophthora black pod disease in Ekiti state of Nigeria: Species differentation and pathogenic variation. Biosciences Biotechnology Research Asia, *5*(1), 157-160. https://doi.org/10.13005/BBRA

[18] Nyassé, S., Grivet, L., Risterucci, A. M., Blaha, G., Berry, D., Lanaud, C., and Despréaux, D. (1999). Diversity of Phytophthora megakarya in Central and West Africa revealed by isozyme and RAPD markers. Mycological Research, *103*(10), 1225-1234. https://doi.org/10.1017/S0953756299008369

[19] Brasier, C. M., and Griffin, M. J. (1979). Taxonomy of 'Phytophthora palmivora on cocoa. Transactions of the British Mycological Society, *72*(1), 111-143.

[20] Offei, S. K., and Oduro, K. A. (1996). Plant pathology in Ghana: Past, present and future. *Ghana Journal of Science*, *36*(1). Volume 31-36, 1991-1996 https://doi.org/10.4314/gjs.v36i1.47992

[21] Adomako, J., Kwoseh, C. K., Emmanuel, M., and Stephen, L. K. (2016). Prevalence of Phytophthora Leaf Blight of Taro (*Colocasia esculenta* (L.) Schott) in the Semi Deciduous Forest Zone of Ghana. *Journal of Experimental Agriculture International*, 1-7. ISSN: 2231-0606

[22] Miyasaka, S. C., Lamour, K., Shintaku, M., Shrestha, S., and Uchida, J. (2013). Taro leaf blight caused by Phytophthora colocasiae. In *Phytophthora a global perspective* (pp. 104-112). Boston, MA, USA: CABI. http://dx.doi.org/10.1079/9781780640938.0104

[23] Omane, E., Oduro, K. A., Cornelius, E. W., Akrofi, A. Y., Asare, E. K., and Barnabas, A. A. (2020). Improvement in the Management of Leaf Blight Disease of Taro (Colocasia esculenta) with Copper and Metalaxyl Fungicides. Journal of Experimental Agriculture International, 127-136. https://doi.org/10.9734/jeai/2020/v42i330490

[24] Maheshwari, S. K., Mishra, R. S., Sriram, S., and Sahu, A. K. (2007). Evaluation of Colocasia Cultivars against Phytophthora Leaf Blight. Annals of Plant Protection Sciences, *15*(1), 264-265.

[25] Van der Puije, G. C., Ackah, F. K., and Moses, E. (2015). Prevalence of leaf blight disease caused by Phytophthora colocasiae in taro in the Aowin Suaman Districts of Ghana. *Hort Flora Research Spectrum 4 (3): 282, 284*.

[26] Ackah, F. K., van der Puije, G. C., and Moses, E. (2014). First evaluation of taro (*Colocasia esculenta*) genotypes against leaf blight (Phytophthora colocasiae) in Ghana. *HortFlora Research Spectrum*, *3*(4), 301-309. ISSN: 2250-2823

[27] Shakywar, R. C., Pathak, S. P., Kumar, S., and Singh, A. K. (2012). Evaluation of fungicides and plant extracts (botanicals) against Phytophthora colocasiae raciborski causing leaf blight of Taro. J. Plant Dis. Sci, *7*, 197-200.

[28] Omane, E., Oduro, K. A., Cornelius, E. W., Opoku, I. Y., Akrofi, A. Y., Sharma, K. and Bandyopadhyay, R. (2012). First report of leaf blight of taro (Colocasia esculenta) caused by

Phytophthora colocasiae in Ghana. Plant disease, *96*(2), 292. https://doi.org/10.1094/pdis-09-11-0789

[29] Opoku, I. Y., Akrofi, A. Y., and Appiah, A. A. (2002). Shade trees are alternative hosts of the cocoa pathogen Phytophthora megakarya. Crop Protection, *21*(8), 629-634. https://doi.org/10.1016/S0261-2194(02)00013-3

[30] Opoku, I.Y. 1994. Survival of Phytophthora palmivora and Phytophthora megakarya in soil and in cocoa roots. PhD. Thesis, University of London, UK.

[31] Abdulai, M., Norshie, P. M., and Santo, K. G. (2020). Incidence and severity of taro (*Colocasia esculenta* L.) blight disease caused by Phytophthora colocasiae in the Bono Region of Ghana. *SSRG International Journal of Agriculture and Environmental Science*, *7*(2), 52-63. ISSN: 2394 - 2568

[32] Ellis, D. H., Davis, S., Alexiou, H., Handke, R., and Bartley, R. (2007). *Descriptions of medical fungi* (pp. 61-167). Adelaide: University of Adelaide.

[33] Ellis, D. H., Merz, W. G., and Hay, R. J. (2005). Topley and Wilson's Microbiology and Microbial Infections; Medical Mycology.

[34] Asare-Nyako, A., and Dakwa, J. T. (1974). The disease on roots. *Phytophthora disease of cocoa. Ed. PH Gregory. London: Longman*.

[35] Akrofi, A. Y., Appiah, A. A., and Opoku, I. Y. (2003). Management of Phytophthora pod rot disease on cocoa farms in Ghana. Crop protection, *22*(3), 469-477.

[36] Opoku, I. Y., Appiah, A. A., Akrofi, A. Y., and Owusu, G. K. (2000). Phytophthora megakarya: a potential threat to the cocoa industry in Ghana. Ghana Journal of Agricultural Science, *33*(2), 237-248. https://doi.org/10.4314/gjas.v33i2.1876

[37] Adomako, J., Kwoseh, C., Moses, E., and Prempeh, R. N. (2018). Variations in morphological and molecular characteristics of Phytophthora colocasiae population causing leaf blight of taro in Ghana. Archives of Phytopathology and Plant Protection, *51*(19-20), 1009-1021. https://doi.org/10.1080/03235408.2018.1550846

[38] Bandyopadhyay, R., Sharma, K., Onyeka, T. J., Aregbesola, A., and Kumar, P. L. (2011). First Report of Taro (Colocasia esculenta) Leaf Blight Caused by Phytophthora colocasiae in Nigeria. Plant disease, *95*(5), 618. https://doi.org/10.1094/pdis-12-10-0890

[39] Brooks, F. E. (2008). Detached-leaf bioassay for evaluating taro resistance to *Phytophthora colocasiae*. Plant disease, *92*(1), 126-131.

[40] Singh, D., Jackson, G., Hunter, D., Fullerton, R., Lebot, V., Taylor, M., ... and Tyson, J. (2012). Taro leaf blight—a threat to food security. Agriculture, *2*(3), 182-203. https://doi.org/10.3390/agriculture2030182

[41] Tsopmbeng, G. R., Fontem, D. A., and Yamde, K. F. (2012). Evaluation of culture media for growth and sporulation of Phytophthora colocasiae Racib., causal agent of taro leaf blight. International Journal of Biological and Chemical Sciences, *6*(4), 1566-1573. https://doi.org/10.4314/ijbcs.v6i4.16

[42] Bangmarigu, E. and Qineti, A. (2018). Cocoa Production and Export in Ghana. Paper prepared for presentation for the 162nd EAAE Seminar. The evaluation of new CAP instruments: Lessons learned and the road ahead. April 26-27, 2018

[43] Shahin, A. S., Jonathan, S., Lary, D. J., Strem, M. D., Meinhardt Lyndel, W.

and Bailey Bryan, (2017). Phytophthora megakarya and P. palmivora, Causal Agents of Black Pod Rot, Induce Similar Plant Defense Responses Late during Infection of Susceptible Cacao Pods. Front. Plant Sci., 14 February 2017. https://doi.org/10.3389/fpls.2017.00169

[44] Naminse, E. Y., Fosu, M., and Nongyenge, Y. (2011). The impact of mass spraying programme on cocoa production in Ghana. *Report on Field Survey*. https://www.ifama.org/resources/files/2012-Conference/556_Paper.pdf. Accessed: 2021/06/08

[45] Tyson, J. L., and Fullerton, R. A. (2015). A leaf disc assay for determining resistance of taro to Phytophthora colocasiae. New Zealand Plant Protection, *68*, 415-419. https://doi.org/10.30843/nzpp.2015.68.5821

[46] Nelson, S., Brooks, F., and Teves, G. (2011). Taro leaf blight in Hawaii.

[47] Asumadu H., Omenyo E. L. and Tetteh F., (2011). Physiological and economic implications of leaf harvesting on vegetative growth and cormel yield of cocoyam (Xanthosoma sagittifolium). Journal of Agrononmy, 10, 112-117. https://dx.doi.org/10.3923/ja.2011.112.117

[48] Quaye, W., Adofo, K., Agyeman, K. O., and Nimoh, F. (2010). Socioeconomic survey of traditional commercial production of cocoyam and cocoyam leaf. African Journal of Food, Agriculture, Nutrition and Development, *10*(9): 4060-4078.

[49] Sagoe, R. (2006). Climate change and root crop production in Ghana. *A report prepared for Environmental Protection Agency (EPA), Accra-Ghana February*.

[50] Sagoe, R., Haleegoah, J., and Marfo, K. A. (2006). Economic evaluation of cocoyam (Xanthosoma sp.) and maize (Zea mays) in a mixed cropping system. *Ghana* Journal of Agricultural Science, *39*(2), 171-180. https://doi.org/10.4314/gjas.v39i2.2140

[51] Iwaro, A. D., Sreenivasan, T. N., and Umaharan, P. (1997). Phytophthora resistance in cacao (Theobroma cacao): Influence of pod morphological characteristics. Plant Pathology, *46*(4), 557-565. https://doi.org/10.1046/j.1365-3059.1997.d01-47.x

[52] Akrofi, A. Y., Opoku, I. Y., and Appiah, A. A. (1995). On-farm farmer managed trials to control black pod disease caused by Phytophthora megakarya in Ghana. In *Proceedings of the First International Cocoa Pests and Disease Seminar, Accra, Ghana, November, 1995* (pp. 6-10). https://doi.org/10.1016/S0261-2194(02)00193-X

[53] Mbong, G. A., Fokunang, C. N., Fontem, L. A., Bambot, M. B., and Tembe, E. A. (2013). An overview of Phytophthora colocasiae of cocoyams: A potential economic disease of food security in Cameroon. Discourse Journal of Agriculture and Food Sciences, *1*(9), 140-145. ISSN: 2346-7002

[54] Hunter, D., Pouono, K., and Semisi, S. (1998). The impact of taro leaf blight in the Pacific Islands with special reference to Samoa. Journal of South Pacific Agriculture, *5*(2), 44-56.

[55] Jackson, G. V. H., Gollifer, D. E., and Newhook, F. J. (1980). Studies on the taro leaf blight fungus Phytophthora colocasiae in Solomon Islands: Control by fungicides and spacing. Annals of Applied Biology, *96*(1), 1-10. https://doi.org/10.1111/j.1744-7348.1980.tb04762.x

[56] Padmaja, G., Devi, G. U., Mahalakshmi, B. K., and Sridevi, D. (2017). Characterization of Isolates of Phytophthora colocasiae Collected from Andhra Pradesh and Telangana Causing Leaf Blight of Taro. International Journal of Current Microbiology and Applied Sciences, *6*, 1901-1912. https://doi.org/10.20546/ijcmas.2017.610.229

Chapter 6

Recognition and Early Stage Detection of *Phytophthora* in a Crop Farm Using IoT

Pooja Vajpayee and Kuldeep Kr. Yogi

Abstract

Detection of agricultural plant pests is seen as one of the farmers' problems. Automated Pest Detection Machine enables early detection of crop insects with advanced computer vision and image recognition. Innovative research in the field of agriculture has demonstrated a new direction by Internet of Things (IoT). IoT needs to be widely experienced at the early stage, so that it is widely used in different farming applications. It allows farmers increase their crop yield with reduced time and greater precision. For the past decade, climate change and precipitation have been unpredictable. Due to this, many Indian farmers are adopting smart methods for environment known as intelligent farming. Smart farming is an automated and IOT-based information technology (Internet of Things). In all wireless environments IOT is developing quickly and widely. The Internet of Things helps to monitor agricultural crops and thus quickly and effectively increase farmers' income. This paper presents a literature review on IoT devices for recognizing and detecting insects in crop fields. Different types of framework/models are present which are explaining the procedure of insect detection.

Keywords: Internet of things (IoT), smart agriculture, pest detection, deep learning

1. Introduction

The livelihoods of Indians are mainly from agriculture. It has been noted in the last decade that there has not been much agricultural crop development. As crop prices decrease, food prices are constantly increasing. Since 2010 more than 40 million people have been driven into poverty [1]. This may be due to water wastes, low soil fertility, abuse of fertiliser, climate change and diseases, etc. There are numerous factors responsible Effective farm intervention is very important and IOT is the solution for integration with wireless sensor networks. It is capable of changing the way agriculture develops and contributes greatly to making smart agriculture. There is a three-tier system in the internet. It contains the layer, network layer and application layer of perception. Sensor motes include perception layer. Devices enabled by ICT, sensor motes are the building blocks of sensor technology. It comprises cameras, RFID tags, sensors and network sensors for object recognition and real-time information collection. The network layer is a universal service IOT infrastructure. The combination of the layer of perception and the application layer is directed. The layer of application is a layer that combines the IOT with specific industry technology.

The internet has almost been applicable in all industries, including intelligent agriculture, smart parking, environmental monitoring for intelligent buildings, health transport and much more.

1.1 Internet of Things (IOT)

The Internet of Things (IOT) is the easiest and most powerful way to solving problems. IOT is established from different assemblies, with tonners of recorders, software, pieces of axes. It also makes details more detailed. Without human interference IOT permits the sharing of data over a network. We will mirror things naturally on the internet as everyday people like a sensor, a car driver etc. An IP address is given so that data can be transmitted across a network. In 2016, the number of connected devices grew 30% relative to 2015 according to the report produced by Garner. He adds that this number is set to grow by 26 billion by 2020 [2].

For the following factors, IOT technology is more efficient:

- Faster Access
- Efficient Communication
- Time Efficiency
- Global Connectivity through any devices
- Minimum human efforts

2. A model for smart agriculture using IoT

In 2016, Patil and Kale [3] reviewed climate change and rainfall over the past decade as annoying. As a result, many Indian farmers are implementing climate-intelligent practices in recent times called intelligent agriculture. Smart farming is an integrated and IOT-driven information technology (Internet of Things). In all wireless settings IOT evolves easily and broadly. Within this article, the introduction of sensor systems and wireless communications to the implementation of IOT technologies and the systems in current agriculture are analyzed and evaluated, in conjunction with the current conditions of the agricultural system Remote Management System (RMS) is proposed to incorporate an approach with internet and cellular communications. Main aim is to capture real-time farming production data offering easy access to farming equipment such as fast massage service warnings and guidance on weather conditions, crops etc (**Figure 1**).

3. Design and realization of a real-time detection device for insect pests of field crops

Mercedes, S., Bo, G. And Yuxia, H., in 2011 [4] investigated the vast species and huge quantity of insect parasites of field crops. Hundreds of common insect pests in farmland are caught by lamps. After the Black Light was trapped, the insect pests were manually recognized and numbered. And the process of predicting was used primarily for a long time in China. It was closely linked to the overall quality of the forecast person's impact, accuracy and efficacy, and was ultimately determined by greater subjective factors.

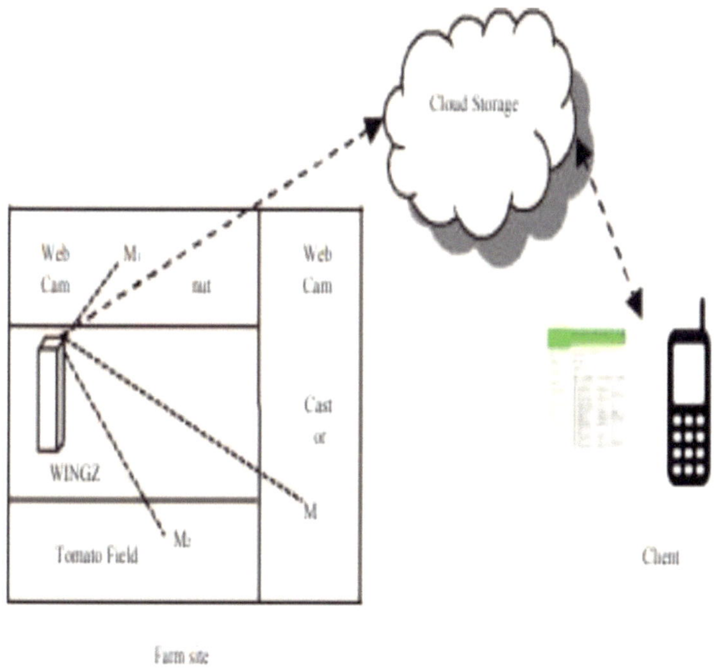

Figure 1.
SAIoT model.

Students studied the picture identification on field plants of insect pests in detail. The static digital camera pictures were categorized of 40 species (25 families, 8 orders) of nightmarish insect pests [5]. With gray-scale images, the 30-five typical plagues (Lepidoptera) manually put were recognized [6]. The auto trapped, killed, and placed into laboratory for further study the eight species of insect pests in cotton fields [7].

3.1 Hardware and software system

With the hardware and software framework the second-generation insect pest detection device was developed. Hardware provided trapping, astonishment and buffering, a uniform lighting, a dispersion-transporter and a vision machine. The software framework has improved image, segmented images, chosen functionality and known insect pests. The unit carried out the entire automation from the collection of insect pests to the identification.

3.2 Hardware design of the detection system

See **Figure 2**.

3.3 Software design of the detection system

Owing to the vast number of details and high efficiency in real time, the cost of the image processing equipment has been raised. Windows 2003 used the framework and the Visual C++ 6.0 language for visual development. MFC-based application software was developed using an OK C30S acquisition graphics card with API functions.

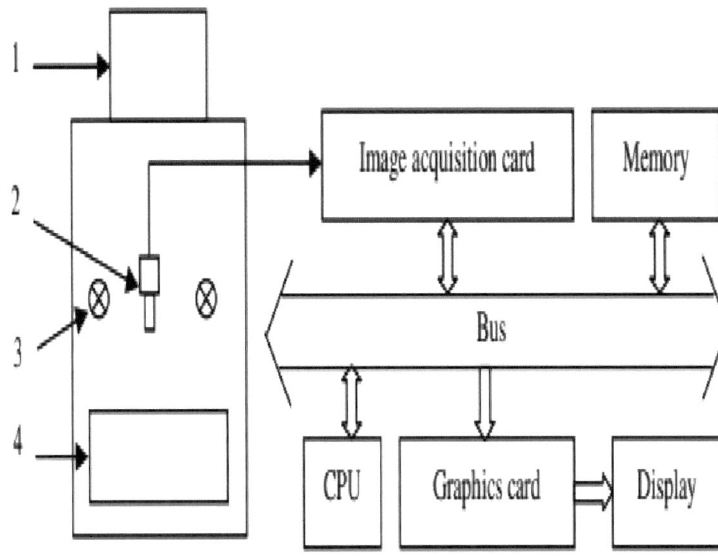

Figure 2.
Hardware components for real time monitoring system for insect pests on field crops. (1) "Trapping, stunning and buffering unit, (2) CCD (3) Illumination unit and (4) Scattering and transporting unit".

3.4 Inference

Real-time detection device for field plagues of second generation has been developed. It performed all automation, from the capture, dispersion, transport, collecting of images, picture analysis to pest recognition. It reduced the duration of detecting pests and increased the degree of automation. The right identification ratio of nine species of pests surpassed 86%. The study further focused on enhancing pest species and improving the detection efficiency.

4. Internet of Things application to monitoring plant disease and insect pests

Shi, Y., Z., Wang, X. Shi, Y. And Zhang, S. in 2015 [8] studied the efficient way of enhancing agricultural low-tech culture quality by using information and communication technology to establish a plant disease and long-term insect pest control system as the farm expert was unable to use the farm for farm management and insecticide disease control. The article introduces internet-based information perception technology (IOT) as well as the role of IOT technology for agricultural disorders and for the control of insect pests, including farming disabilities and the insect pest control system, the collection of sensor nodes, data processing and exploitation of insect pest information, etc. An IOT-based disease and insect pest control system consisting of three levels and three systems were proposed. A new way of accessing agricultural information on the farm is provided by the system.

5. Agricultural crop monitoring using IoT

"Sreekantha, D.K. and Kavya, A.M", in 2017 [9] investigated the reorganisation of the IOT for agriculture, helping farmers to deal with problems in the region with a broad variety of technology including accuracy and sustainable agriculture. IOT technology helps to gather information on conditions such as weather, precipitation, soil temperatures and fertility, field tracking online allows weed identification, water level, pest detection and interference of livestock, crops growth and farming. IOT moves farmers from everywhere and at any time to link to their farm. Farms are monitored using wireless sensor networks, and microcontrollers for the control and automation of agricultural processes. Wireless cameras have been used to view the environments remotely in the form of pictures and images. A smart telephone enables farmers, at every time and anywhere in the world to keep up to date with the current conditions of their agriculture. IOT technology can cut costs and increase conventional agriculture's productivity.

5.1 Inference

The Internet of things helps agricultural crop tracking to increase crop production and thus the farmers' income quickly and effectively. A wireless sensor network and sensors of all kinds are used for the collection of crop conditions and environmental change and are transmitted to farmers/devices via a network to cause corrective action. Farmers are in connection with the circumstances of the agriculture sector around the globe at any moment and wherever. Few connectivity drawbacks have to be addressed by encouraging the technologies to conserve resources and also by facilitating the user experience.

6. Remote insects trap monitoring system using deep learning framework and IoT

"Ramalingam, B., Mohan, R.E., Pookkuttath, S., Gómez, B.F., Sairam Borusu, C.S.C., Wee Teng, T. and Tamilselvam, Y.K in 2020" [10] researched that early insect identification and control (human physical conditions as an example houses, hospitals hotels, parks, camps, flooring, industries related to food etc) and agricultural farms were important for developed environments. These pest control steps are currently labor-dependent manual, repetitive, unpredictable and time-consuming activities. Latest advances in Internet of Things (IOT) and Artificial Intelligence (AI) and the can automate a range of maintenance operations, improving efficiency and safety dramatically. This document includes the implementation of Deep Learning (DL) and IOT monitoring system of insect traps in real-time as well as the detection of insects. "*The system architecture for remote trap monitoring is developed with IOT and the unified target detection framework of faster RCNN (Region-based convolution neural networks) Residual neural Networks 50 (ResNet50). The object detection system for Faster RCNN (ResNet 50) was trained and deployed in IOT using designed environmental insects and farm insect imagery*". The proposed device was tested with four-layer IOT and the picture of constructed ecosystem insects caught by sticky trap sheets in real time. In addition, insects from farms have been examined by a different database of photographs of insects (**Figure 3**).

6.1 Inference

In this article, the IOT and deep learning system was proposed for the remote insect tracking and the automated process of insect detection. The Faster version of RCNN ResNet object identification mechanism was accustomed to automately classifies the parasite type, using the Four-Layer IOT system to built the remote trap insect tracking mechanism. Included is an ecosystem insect data base and farm field insect archive that has been checked for offline and on-line reliability in

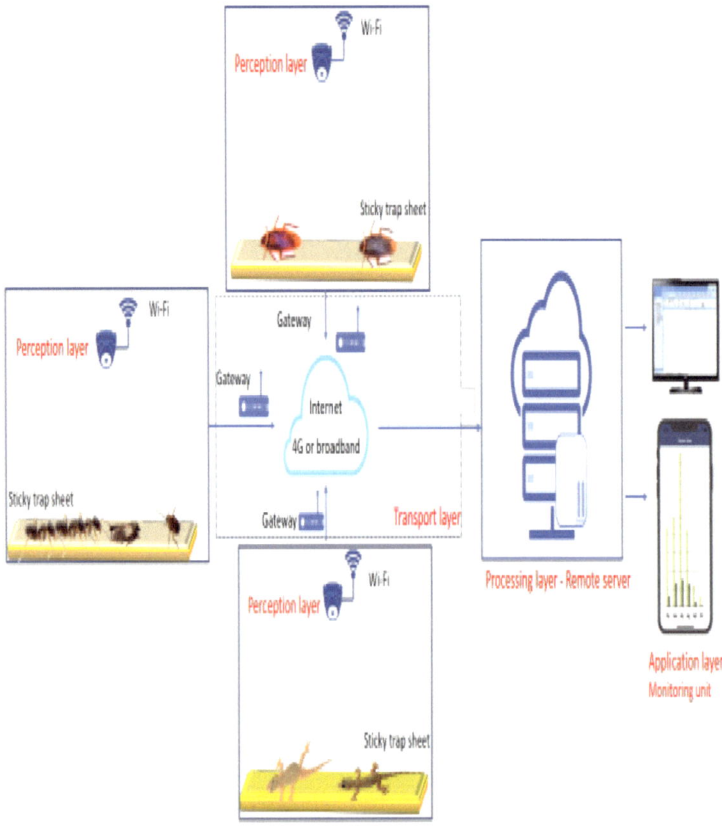

Figure 3.
System for remote trap control and insect detection based on IOT and DL.

the detection of intensely learning insects. According to other object recognition mechanisms such as SSD and Yolo, the accepted device provided optimal insect detection efficiency. The research has shown that 96 percent of insects identified with built-in environmental insects were obtained by the qualified model, 94 percent were identified with farmland insects, and 0.2 s were needed on average for processing the one image. This case study has demonstrated the automation of remote identification through IOT and a DL-based insect monitoring system using a qualified CNN framework and overcomes insect control systems failure.

7. Automated remote insect surveillance at a global scale and the Internet of Things

"Potamitis, I., Eliopoulos, P. and Rigakis, I" in 2017 [11]. In many of our records, a large number of extreme insect plagues of agricultural and health importance were studied in a broad spatial scale as the principle of remote insect control. The trap is used to make the trap, the timeline, the GPS tag and where necessary, the inbound insect species from the wing beat safe to inject. Standard low-cost pest traps for certain insect species augment the insects. Both large crop insects are tracked in order to decide if a treatment strategy can be undertaken before a significant infestation takes place. Monitoring processes are based on specially designed mosquito traps. Conventional insect monitoring is used in the spectrum of such tracking. It takes physical labour, consumes money and also needs an expert to be accurate enough, often with the potential for raising human security concerns. It is limited on its own expenses.

These drawbacks decrease the amount of manual insect surveillance and thus its precision, which eventually contributes to considerable cultivation losses due to pest damage. You intend to monitor the existence, the stamping period, the detection of species and the population density of target plaguicides with 'supervision' to unmatched data extraction levels. Insect counts are wirelessly transmitted to the central supervisory agency, and predictive tools for the control of insect pests and environmental factors related to population growth are visualized and streamed through. The work illustrates how traps can be organized in networks that report collectively on local, state, world, continental and international data using the new Internet of Things technologies (IOT).

This research is undoubtedly cross-disciplinary, lies at the intersection of entomology, optoelectronic engineering and computer and crop-science and involves the production and introduction by many of the most important agricultural pests of low-cost low-energy technologies to minimize the amount of quantitative and qualitative crop failure. They claims that clever traps interacting through IOT will have a huge effect on the crops security decision-making process in real time and would undermine current manual practices directly from the field to a manually managed entity in the very near future. In this article three instances of Rhynchophorus ferrugineus are investigated: track the use of (a) picusan, (b) lindgren and (c) monitoring various stored grain beetle pests using sored grain pitfall trap for Rhynchophorus ferruginus (Olivier) (Coleoptera: curculionidae). There is a very detailed approach to the industry that delivers accuracy per cent on automated quantity when as opposed to the actual defined number of insects in each type of trap.

8. IoT monitoring system for early detection of agricultural pests and diseases

"Materne, N. and Inoue, M. in 2018" [12] examined the advancement in sensor technology has driven the technological revolution in agriculture. In these days, however the opportunity to use digital technologies relating to the Internet of Things (IOT) increase sharply; the growth of the roles system of farming continues in its early stages. The current threats of less preferred climate conditions thrive on the increased risk of cross-border plant pesticides and diseases that damage crops, as well as on the danger to the food security and on some major losses for the farmers, provided that agricultural sector is still suffering from climate changes. In this study, they merged Wireless Sensor Network (WSN) sensor devices to establish an agricultural field monitoring framework that simultaneously tracks eight primary environmental parameters that are known as highly interconnectioned to booming pests and plant diseases. The overall system configuration provided for real-time tracking and regular collection of the huge volume of data. This is why they have investigated the knowledge obtained using machine learning approaches using "KNN, Random Forest, Logistic Recovery and Linear Regression algorithms". The purpose of this article is to perform an experiment on the advantage of using the IOT systems in agricultural lands to gather and analyze data to determine a prediction model that could be used to help forecast the outbreaks of plantation diseases.

8.1 Inference

They suggested an IOT framework with functionalities for day-to-day tracking of farmland environmental parameters. They also developed a predictive model for the provision by applying machine learning algorithms to avoid the outbreaks of pest and diseases in planting. The study merged IOT and machine learning technologies to improve agriculture and agriculture to draw on new concepts and developments in technology in order to sustain, increase yields and increase agricultural efficiency. The work represents only eight sensed parameters; the number of sensors, like meteorological information, may be added for potential production purposes. In addition, it is important to enhance the work of cloud providers so long as the behaviour and features of each type of pest and disease are taken into account.

9. Research on insect pest image detection and recognition based on bio-inspired methods

"Deng, L., Wang, Y., Han, Z. and Yu, R" in 2018 [13] performed a study entitled "Research on insect pest image detection and recognition based on bio-inspired methods". In this study, methods inspired by human visual systems were suggested to easily identify and identify insect pests. SUN was used for the generation of saliency and area of interest maps and identification in pest images using a Natural Statistics Model (NGM) to inspire human visual focus. In order to exclude invariant characteristics, the bio- influenced Hierarchic Model and X (HMAX) model were used to reflect a plague presence. Scale Invariant Function Transform (SIFT) in the HMAX model was integrated to improve rotational changes invariance. In the meanwhile, Coarse non-negative encoding (NNSC) simulates clear cell replies. Furthermore an Invariant Texture Properties has been removed using the Local Pattern (LCP) algorithm. Finally, the extracted characteristics were provided for recognition by Support Vector Machines (SVM). Experimental studies have shown a gain from the proposed approach relative to the approaches "HMAX, Sparse Coding and Normal Input Memory (NIMBLE), which is similar to the Deep Convolution Network" (**Figure 4**).

9.1 Framework

See **Figure 5**.

9.2 Result

With an identification rate of 85.5%, the proposed approach obtained a successful outcome and could efficiently identify insect pests in diverse ecosystems. The suggested solution offered a new approach to the identification and recognition of insect pests.

"P. Tirelli, N.A. Borghese" [14] found that surveillance of the population of pesticides in the plant sector is currently a concern. At present, the device is based on dispersed images, which can be automatically collected and transmitted images of stuck areas to a remote station by means of a wireless sensor network. The station tests the density of the production of insects at various farm sites and alerts when the threshold of the insect is surpassed. The client nodes are spread in the fields that serve as monitoring stations. The main node co-ordinates the network and retrieves images from client nodes. During the four week monitoring cycle, the network periodically operates, and the viability is assessed, forecasts the population curve of the pest insects compared with everyday assessment.

Figure 4.
Sample images taken in natural conditions. (a) Locusta migratoria, (b) Parasa lepida, (c) Gypsy moth larva, (d) Empoasca flavescens, (e) Spodoptera exigua, (f) Chrysochus chinensis, (g) Laspeyresia pomonella larva, (h) Spodoptera exigua larva, (i) Atractomorpha sinensis, (j) Laspeyresia pomonella.

10. Insect pest image detection and recognition based on bio-inspired methods

"Nanni, L., Maguolo, G. and Pancino, F" in 2020 [15] performed a study entitled "Insect pest image detection and recognition based on bio-inspired methods". Identification of insect pests is important for crop safety in many areas of the world. They introduce in this article an artificial classification based on the combination of saliency and neural networks. SALIENCY techniques are widespread image processing algorithms that identify the most important image pixels. In this paper you are using three different salience approaches as preprocessing of images and make three distinct images for each form of saliency. For each original image they generate new photographs to train various neural convolution networks, they create 3*3 = 9 new pictures (**Figure 6**).

10.1 Result

They evaluate each execution of each preprocessing/network pair as well as they examine the performance of their grouping. You assess your solution on the major IP102 and a tiny dataset. Its best ensembles achieve the maximum degree of

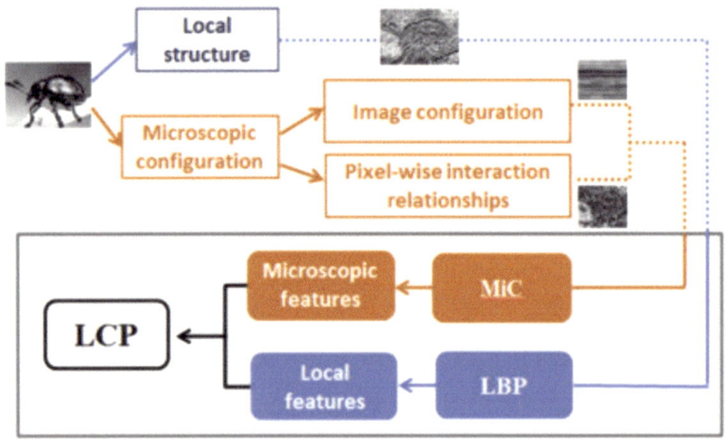

Figure 5.
Feature extraction framework of LCP.

Figure 6.
Image samples from IP102.

Species	Number of Samples
Locusta migratoria	72
Parasa lepida	59
Gypsy moth larva	40
Empoasca flavescens	41
Spodoptera exigua	68
Chrysocus chinensis	50
Laspeyresia pomonella larva	50
Spodoptera exigua larva	56
Atractomorpha sinensis	62
Laspeyresia pomonella	65

Table 1.
Composition of dataset.

technical precision, with both the smaller (92.43%) and the IP102 (61.93%), matching the efficiency of smaller human experts (**Table 1**).

11. Automatic detection and monitoring of insect pests-A review

Cardim et al. in 2020 [16] studied that certain insect pest species can be automatically identified and tracked. In order to develop integrated pesticide management (IPM) for precision agriculture, several systems were developed. For many essential pests, automatic detection traps were developed. This emerging methods and strategies are very promising to identify hostile and quarantine pests early on and to monitor them. The paper attempts to evaluate technological and scientific state-of-the-art sensor technologies in order to track and automatically identify insect pests (**Figure 7**).

"In the article the methods are discussed for the identification, the applications are introduced, as well as recent progress, comprising machine learning and the Internet of Things, infrared monitors for pests, audio sensors and classification through images".

Figure 7.
Automatic trap for moth species tracking (a) and fruit flies (b) EFOS, Trapview, Slovenia.

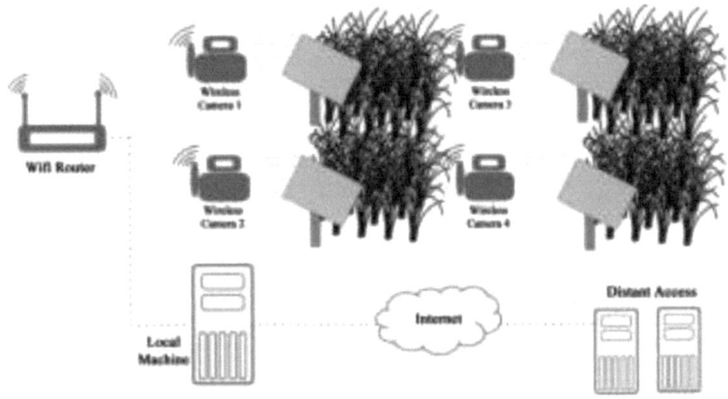

Figure 8.
Global architectural design.

12. Pest detection and extraction using image processing techniques

"Miranda, J.L., Gerardo, B.D. and Tanguilig III, B.T" in 2014 [17] performed a study entitled "Pest Detection and Extraction Using Image Processing Techniques". Detecting pests in paddy fields is an important problem in agriculture, so effective steps to tackle infestation and reduce the use of pesticides should be created. Imaging techniques are commonly used in agriculture to offer optimum protection for crops, thereby leading to greater crop management and production. The surveillance of pests depends on workers but electronic control is developing to mitigate the efforts and mistakes of human beings (**Figure 8**).

This study broadens the application of various imaging procedures for the identification and extraction of insect pests through the establishment of an integrated paddy field detection and extraction device for estimating plague densities. Experimental review shows the proposed approach to detect pests in rice fields to provide a quick, reliable and simple solution.

13. Insect detection and classification based on an improved convolutional neural network

"Xia, D., Chen, P., Wang, B., Zhang, J. and Xie, C" in 2018 [18], to solve the issue of multi-classification of insects in the field, evaluated a prototype of a neural network. The model will use the benefits of the neural network such that multifaceted insect traits are completely extracted. In the regional proposal process, rather than a standard, selection technique, the regional proposal network is implemented in order to produce fewer proposal windows, which are particularly valuable for improving the accuracy of forecasts and accelerating calculations. Experimental studies indicate that the proposed approach is better than the current conventional insect classification algorithms and is higher reliably (**Figure 9**).

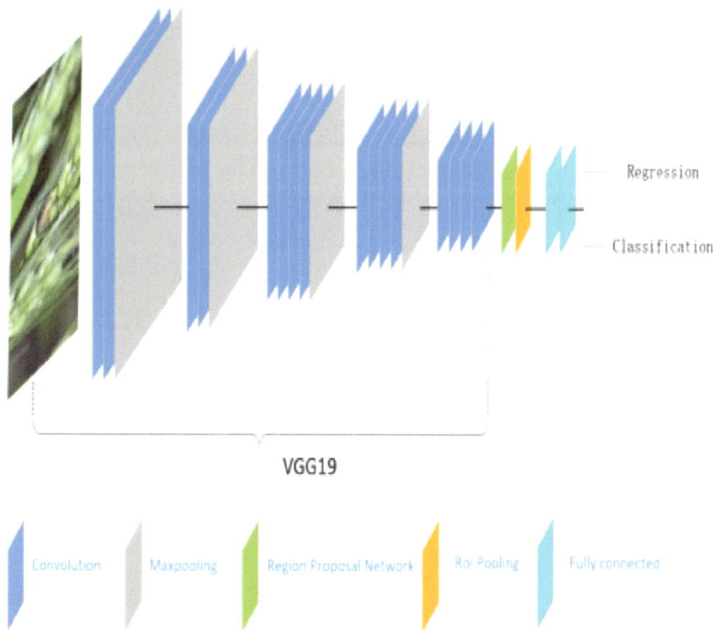

Figure 9.
The proposed detection model using VGG19 having schematic structure.

14. Image processing techniques for insect shape detection in field crops

Thenmozhi, K. and Reddy, U.S., 2017 [19] performed a study entitiled "Image Processing Techniques for Insect Shape Detection in Field Crops". In farming, the identification of crop pest is considered one of the farmers' challenges. An integrated machine vision and picture analysis device for insect detection enables improved identification of early-stage plant insects with a lower time and greater precision that will improve crop yield. Digital image processing methods were used in the current work to detect the insect forms in the sugar cane crop with photographs of crop insects for pre-processing, segmentation and extraction. Sobel edge detecting is introduced to distal media. In the extraction of the feature nine geometrical features can be defined in the structure of the insect. This recognition of insect shape performs well and achieves high precision for round (circular), oval, triangular and rectangular sugarcane field insects. The study was carried out using the Image Processing Toolbox in MATLAB 2015b (**Figure 10**).

15. IOT-based drone for improvement of crop quality in agricultural field

"Saha, A.K., Saha, J., Ray, R., Sircar, S., Dutta, S., Chattopadhyay, S.P. and Saha, H.N.", in 2018 [20] researched that the need for increased population and agriculture is becoming increasingly frequent with unmanned air vehicles. Drones with suitable cameras, sensors and modules can help to make agriculture simpler, more effective and more accurate. The solutions proposed relating to these drones will help to expand the potential of further improvement if combined with various machine learning and the Internet of Things concepts. The relevant work in

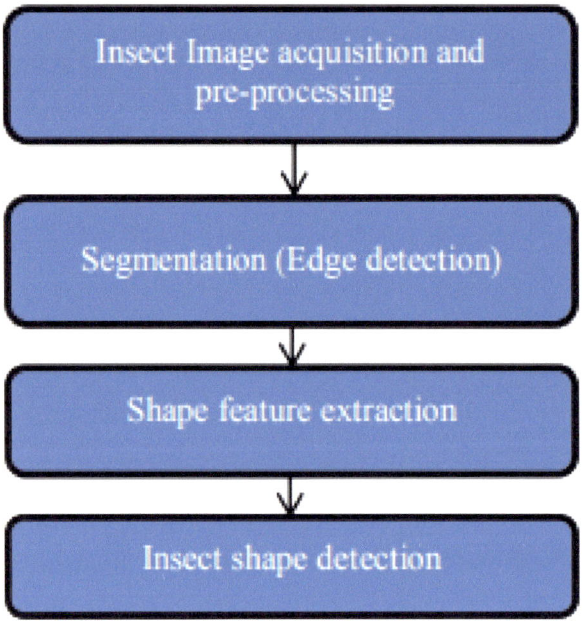

Figure 10.
Flow chart for detecting crop insects form.

this field and the solutions which could be incorporated into the drone using the Raspberry Pi 3 B module were highlighted in this paper.

15.1 Inference

They conclude that drones or UAVs would be of tremendous help in agriculture as they are crucial at the start of a crop cycle with an increasing population. Not only will it minimize time, but it will also generate better cultivation based on analyzed data. The systematic monitoring will make crop management more effective. With the next developments, with less electricity usage, the output rate will increase rapidly. Drones are used in the planting of plant nutrients in the soil not just in soil and field analysis but also in planting seeds. The use of drones could also eliminate the crop monitoring obstacles that had previously been faced. Drones are not stopped here because they are integrated into hyper-spectral, thermal, or multi-spectral sensors and drones may detect which parts of the soil are dry. Furthermore, drones will also be used for scanning with near-infrared and visible light in order to determine crop health. Drones therefore act as an ideal aeroplane for the accuracy data collection.

16. In intelligent agriculture, a vision-based flying insects counting and recognition system

"Zhong, Y., Gao, J., Lei, Q. and Zhou, Y" in 2018 [21] studied that a Design and implementation of flying insects counting and classification system based on vision. The system is built as follows: firstly, in the surveillance area a yellow sticky trap is installed to catch flying insects, and the camera is mounted to capture images on the spot. Then the process of detection and

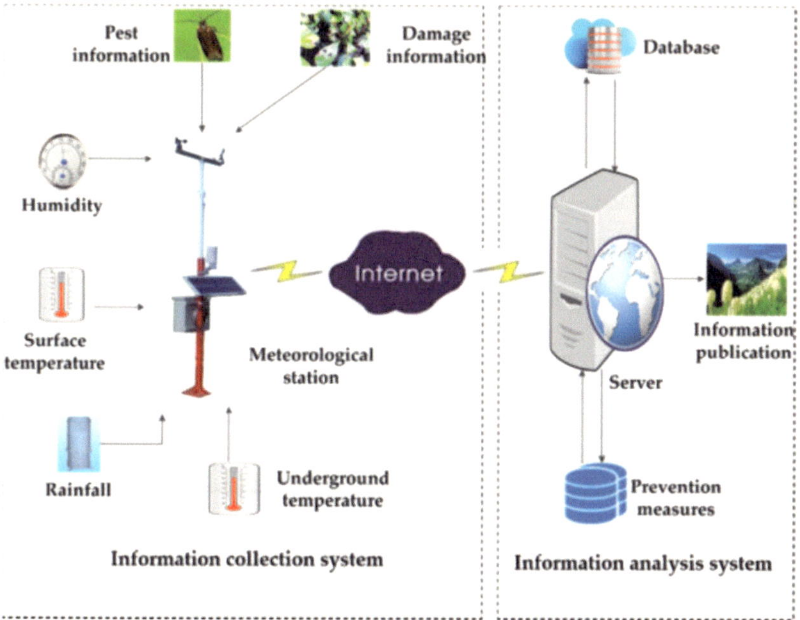

Figure 11.
The components of agricultural monitoring service platform.

coarse counting on the basis of the detection and classification of objects by You Only Look Once (YOLO), using global functionalities is designed. Six insect species including fly, fruit fly, bee, mouth, moth, and chaff have been chosen to evaluate the efficacy of the system. The Raspberry PI insect counting and recognition system is implemented. The results of the tests show promising performance compared to conventional methods. On Raspberry PI the average counting accuracy is 92.5% and average classification accuracy is 90.18 percent. The system proposed is simple to use and provides fast and reliable identification data; it can therefore be applied for smart farming applications (**Figure 11**).

17. Pest24: a large-scale very small object data set of agricultural pests for multi-target detection

"Wang, Q.J., Zhang, S.Y., Dong, S.F., Zhang, G.C., Yang, J., Li, R. and Wang, H.Q.", in 2020 [22] studied that Accurate agriculture poses new challenges for on-site pest monitoring in real time based on the new AI technology generation. This paper establishes a large-scale standardised data collection of agricultural pests, called Pest24 to provide a large data resource for the training of profound learning models for the detection of pests. In particular, the current data set consists of 25.378 images from our automatic plague trap and imaging system annotated with pests. 24 typical pest categories, which are mainly responsible for destroying field crops in China every year, are involved in Pest24. They use various cutting-edge methods of detection, such as RCNN Faster, SSD, YOLOv3 and Cascade R-CNN, to detect pests in the data set and to generate promising findings for in real time field pests monitoring. In the exploration of factors that impact on pest detection accuracy, the data set are analysed into a range of aspects and the three factors that mainly affect the performance of the pest detection, i.e. relative scale, number of instances and adhesion of objects. Overall, Pest24 usually features large-scale multi-pest image data, small object sizes, high object resemblance and dense pesticide distribution.

Numbers of images and instances of each pest category in Pest24

Index	Pest name	Images	Instances
1	Rice planthopper	316	1511
2	Rice Leaf Roller	944	1240
3	Chilo suppressalis	454	1285
5	Armyworm	3828	8880
6	Bollworm	9049	28,014
7	Meadow borer	5526	16,516
8	Athetis lepigone	7520	30,339
10	Spodoptera litura	1588	1951
11	Spodoptera exigua	3614	7263
12	Stem borer	1357	1804
13	Little Gecko	2503	4279
14	Plutella xylostella	531	953
15	Spodoptera cabbage	1707	2302
16	Scotogramma trifolii Rottemberg	3223	4679
24	Yellow tiger	1388	1686
25	Land tiger	369	475
28	Eight-character tiger	154	168
29	Holotrichia oblita	90	108
31	Holotrichia parallela	3111	11,675
32	Anomala corpulenta	5228	53,347
34	Gryllotalpa orientalis	3629	6528
35	Nematode trench	118	167
36	Agriotes fuscicollis Miwa	1814	6484
37	Melahotus	239	768

18. Internet of things for smart agriculture: technologies, practices and future direction

"Ray, P.P" in 2017 [23] studied that innovative science in agriculture has taken a new path with the introduction of the Internet of Things (IoT). IoT must be widely experimented to be applied in different agricultural applications at the emerging stage. In this paper, they review many possible IoT applications and the particular questions and challenges relating to improved agricultural IoT deployment. The devices and wireless communication technologies linked to IoT are evaluated extensively to meet the unique requirements of agricultural and agriculture applications. Research has been carried out on those IoT sensor systems which deliver intelligent and intelligent services to smart farming. Various case studies are provided to analyse the IoT-based solutions implemented according to their implementation parameters by different organisations and individuals and categories. The related issues in these solutions are also illustrated, while defining progress factors and potential IoT work maps.

19. Real-time detection of apple leaf diseases using deep learning approach based on improved convolutional neural networks

"Jiang, P., Chen, Y., Liu, B., He, D. and Liang, C.", in 2019 [24] examined that The Alternarial leaf spot, the brown spot, the mosaic, the grey spot and Rust are five common types of apple leaf conditions. However, the detection of apple diseases in current research is not reliable and easy to ensure that the apple industry is growing healthy. This paper provides a deeper learning approach aimed at the detection in real time of enhanced neural networks (CNNs). This article is the first use of the Apple Leaf Disease Data Set (ALDD) data augmentation and annotation technology which consists of laboratory pictures and complex images in real life. Based on this proposal a new apple leaf detection model that uses deep-CNN is proposed by integrating the GoogLeNet Inception and Rainbow concatenation system. Finally, five common Aple Leaf diseases using a dataset of 26,377 image of diseased Apple leaves are trained by the proposed INAR-SSD (SSD with Initiation and Reinbow Concatenation Module) model. The experimental results show that a high rate of 23.13 FPS 78.80 per cent of mAP is used for the INAR-SSD model in the ALDD. The results show that an early diagnosis approach to apple leaf disease has been established by the new INAR-SSD model, which detects these diseases in real time with greater precision and rapidity than before.

20. A review of advanced machine learning methods for the detection of biotic stress in precision crop protection

"Behmann, J., Mahlein, A.K., Rumpf, T., Römer, C. and Plümer, L.", 2015 [25] studied that the early and accurate biotic stress detection is needed for effective crop protection. Important results have been obtained in recent years on the early identification of weeds, herbal diseases and insect pesticides in crops. The findings relate both to the development of optical sensors that are non-invasive high resolution and methods for data processing that can manage signals from the resolution, size and complexity of these sensors. Several machinery learning methods, such as vector machines and neural classification (supervised learning) networks, have been used for precision agriculture (unsupervised learning). These methods can measure both linear and non-linear models and require few statistics. Early detection of plant diseases by the use of supervised or unattended learning methods, and weed detection by the use of formal descriptors, is effective applications. A brief introduction to machine learning that an analysis of its potential to protect precision crops and an overview of instructive examples from several areas.

21. A novel cloud computing based smart farming system for early detection of borer insects in tomatoes

"Rupanagudi, S.R., Ranjani, B.S., Nagaraj, P., Bhat, V.G. and Thippeswamy, G.", 2015 [26], examined that farmers suffer tremendous losses every year as a result of plague infestations and that affects their livelihoods. In this article we will address a new solution to this problem by continuous video processing, cloud computing and robotics monitoring of crops. The paper describes the methods for pesticide detection in one of the most popular tomato fruits in the world. "An insight into how in this project even the notion of the Internet of Things can be conceptualized".

22. Wireless sensor network based automated irrigation and crop field monitoring system

"Nisha, G. and Megala, J." in 2014 [27] studied that wireless Sensor Automatic irrigation system based on a network to maximise agricultural water use. A network of wireless sensors and temperature sensors distributed with solid moisture in the field is the unit. The Zigbee Protocol is used for the management of sensor information and for the control by means of an algorithm with sensor threshold values to a microcontroller irrigation system. A solar panel and a mobile Internet guide are used for the unit. A wireless camera in the field with image processing techniques monitors the area of the disease. The device is low costs and benefits power independence in water-limited geographically isolated areas.

23. A review of recent sensing technologies to detect invertebrates on crops

This study is performed by "Liu, H., Lee, S.H. and Chahl, J.S." in 2017 [28]. In order to detect pesticides more effectively, researchers have developed different technologies. The existing sensing technology, however, is still limited to effective field applications. This review paper is designed for exploring relative technologies and finding a method for the sensing and detection of crop invertebrates such as butterflies, sauté, snails and slugs. Two main areas for the identification and detection of invertebrates were identified: acoustic sensing and vision system (MVS). The acoustic sensor is suitable for detection and identification of soil pests, stored grains and wood, whereas acoustic sensors must usually be fitted to inspection samples causing difficulties for efficient inland applications. MVS has the potential to detect and identify invertebrates in crops in a more efficient and flexible manner. The invertebrate identification technologies have recently been studied in detail with MVS, but the detection of infertile fields is relatively weak. This study summarizes current research deficiencies and discusses possible research directions.

24. IOT based strawberry disease prediction system for smart farming

This study performed by "Kim, S., Lee, M. and Shin, C.", 2018 [29]. In this study cloud-based technology was built to manage information

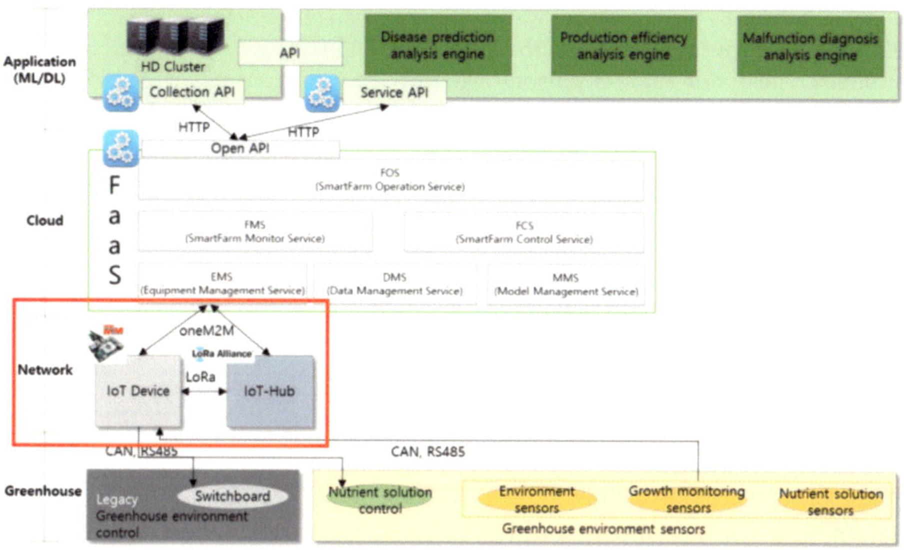

Figure 12.
The Internet of Things (IoT)-Hub is used in this network model. FaaS: Farm as a Service; api: application programming interface; LoRa: Long Range.

gathering, analysis and prediction in a shared framework. The proposed integrated system Farm as a Service (FaaS) assists high-level applications support through farms activity and observing and associated equipment, data and models management. Such a system records, connects and manages IoT devices and analyses information about the environment and growth. Furthermore, this study consisted of the IoT-Hub network model. Effective data transmission for every IoT device and communication for non-standard items is supported by this model and is highly reliable in communication in most difficult circumstances. IoT-Hub thus make sure the steadiness of agricultural environment-specific technology. Specific systems are implemented at different levels in an integrated agricultural specialist FaaS system. A strawberry infection prediction system has been designed and analysed, and this system has been compared with other infection models (**Figure 12**).

25. A multispectral 3-D vision system for invertebrate detection on crops

"Liu, H., Lee, S.H. and Chahl, J.S.", in 2017 [30] examined that the benefits of multi-spectral and hyperspectral vision systems have been demonstrated to detect such invertebrate pests efficiently and accurately. However, the identification of certain camouflaged pests on host plants has been restricted by only use of spectral details. Three-dimensional (3-D) representations are widely studied for multifaceted object recognition and scene perception in many fields. However, because of a lack of appropriate data collection methods and efficient algorithms, 3-D technologies have no invertebrate detection applications. "They created a multi-spectral vision system that enables the development of denser plant and pest clouds with multi-spectral images of UV, blue, green, red, and near-infraround". An algorithm was designed to differentiate wide leafs from relatively large pests in nuclei at the noisy stage. The vision can be used as an automated pesticide sprayer sensor, or to support advanced pesticide monitoring systems.

26. Design of intelligent agriculture management information system based on IoT

Yan-e, D., in 2011 [31] studied that Agricultural IT (AIT) has been the most productive means and instruments for improving agriculture productivity and making maximum use of agricultural resources and has been widely used on all aspects of agricultural. The use of information technology from agricultural production as an important sub technology of AIT measures the level of agricultural computerization and the efficacy of decision-making on farm production. In this paper the methods for MIS intelligent agriculture design and architecture is discussed based on the implementation of the concept of management of agricultural knowledge and analysis of agricultural data characteristics (**Figure 13**).

Early detection of *Phytophthora* spp using IOT and machine learning architecture

Figure 13.
Investigation in Crop Production of Agriculture Information Management Flow.

DOI: http://dx.doi.org/10.5772/intechopen.97767

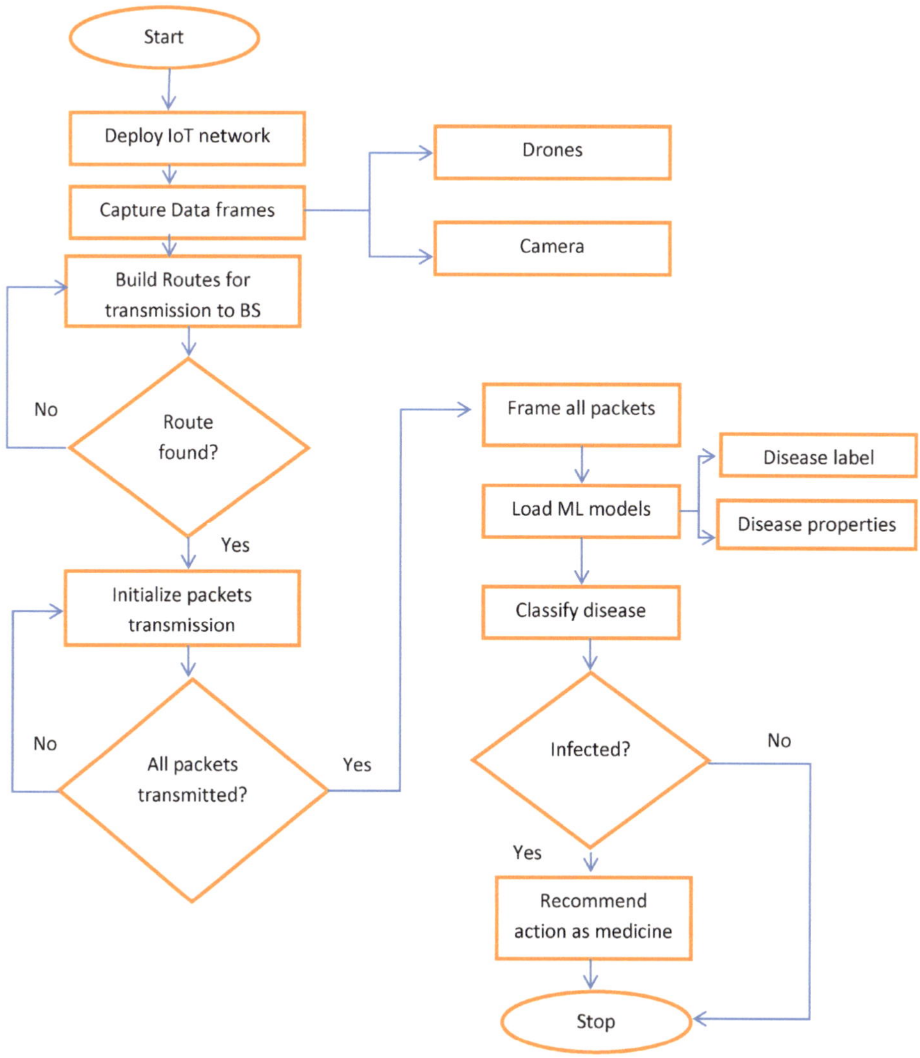

Figure 14.
IoT architecture for Phytophthora *spp. detection.*

See **Figure 14**.

The above flow is a combination of IoT networks and machine learning models to predict and recommend the *Phytophthora* spp. from crops. Here the initial steps to form the architecture are network deployment and configure with initial node parameters. This phase have various sub-modules where the network nodes placed on field, sensing areas formations etc. Once the nodes deployed in the field, network node regularly sense data in the form of captured frames and send it to the base station. The base station having various detection and recommendation models to analyse the data frames.

All transmitted frames from the field to the base station formed in the actual format and load in the ML models. ML models in the network base station are already trained models for various crops disease properties and labels. These are able to classify the captured properties and recommend the disease and various possible treatments. After classification, if the model predicts infection then it suggests the medicines and track for future updates.

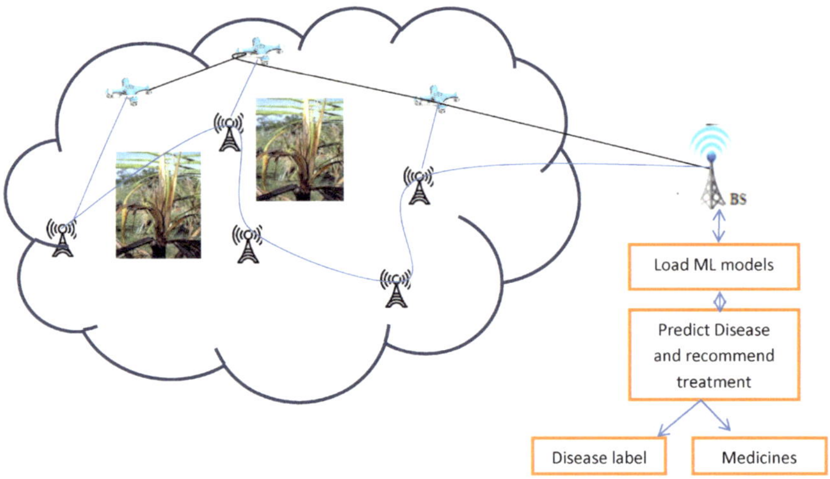

S. No	Author name	Paper description	Publication year
1	Patil and Kale	In all wireless settings IOT evolves easily and broadly. Within this article, the introduction of sensor systems and wireless communications to the implementation of IOT technologies and the systems in current agriculture are analyzed and evaluated, in conjunction with the current conditions of the agricultural system Remote Management System (RMS) is proposed to incorporate an approach with internet and cellular communications.	2016
2	Mercedes, S., Bo, G. And Yuxia, H	Real-time detection device for field plagues of second generation has been developed. It performed all automation, from the capture, dispersion, transport, collecting of images, picture analysis to pest recognition. It reduced the duration of detecting pests and increased the degree of automation.	2011
3	Shi, Y., Z., Wang, X. Shi, Y. And Zhang, S	An IOT-based disease and insect pest control system consisting of three levels and three systems were proposed. A new way of accessing agricultural information on the farm is provided by the system.	2015
4	Sreekantha, D.K. and Kavya, A.M	A smart telephone enables farmers, at every time and anywhere in the world to keep up to date with the current conditions of their agriculture. IOT technology can cut costs and increase conventional agriculture's productivity.	2017
5	Ramalingam, B., Mohan, R.E., Pookkuttath, S., Gómez, B.F., Sairam Borusu, C.S.C., Wee Teng, T. and Tamilselvam, Y.K	Latest advances in Internet of Things (IOT) and Artificial Intelligence (AI) can automate a range of maintenance operations, improving efficiency and safety dramatically. This document includes the implementation of Deep Learning (DL) and IOT prototype for the tracking of insect traps in real-time and the identification of insects.	2020
6	Potamitis, I., Eliopoulos, P. and Rigakis, I	Conventional insect monitoring is used in the spectrum of such tracking. It takes physical labour, consumes money and also needs an expert to be accurate enough, often with the potential for raising human security concerns. It is limited on its own expenses.	2017

S. No	Author name	Paper description	Publication year
7	Materne, N. and Inoue, M.	IOT framework with functionalities for day-to-day tracking of farmland environmental parameters. They also developed a predictive model for the provision by applying machine learning algorithms to avoid the outbreaks of pest and diseases in planting.	2018
8	Deng, L., Wang, Y., Han, Z. and Yu, R	To exclude invariant characteristics, the bio-influenced Hierarchic Model and X (HMAX) model were used to reflect a plague presence. Scale Invariant Function Transform (SIFT) in the HMAX model was integrated to improve rotational changes invariance. In the meanwhile, Coarse non-negative encoding (NNSC) simulates clear cell replies. Furthermore an Invariant Texture Properties has been removed using the Local Pattern (LCP) algorithm.	2018
9	P. Tirelli, N.A. Borghese	The device is based on dispersed images, which can be automatically collected and transmitted images of stuck areas to a remote station by means of a wireless sensor network. The station tests the density of the production of insects at various farm sites and alerts when the threshold of the insect is surpassed.	2011
10	Nanni, L., Maguolo, G. and Pancino, F	SALIENCY techniques are widespread image processing algorithms that identify the most important image pixels. In this paper you are using three different salience approaches as an image preprocessing and create three different images for each form of saliency.	2020
11	Cardim et al	For many essential pests, automatic detection traps were developed. This emerging methods and strategies are very promising to identify hostile and quarantine pests early on and to monitor them.	2020
12	Miranda, J.L., Gerardo, B.D. and Tanguilig III, B.T	Imaging techniques are commonly used in agriculture to offer optimum protection for crops, thereby leading to greater crop management and production. The surveillance of pests depends on workers but electronic control is developing to mitigate the efforts and mistakes of human beings.	2014
13	Xia, D., Chen, P., Wang, B., Zhang, J. and Xie, C	Experimental studies indicate that the proposed approach is better than the current conventional insect classification algorithms and is higher reliably.	2018
14	Thenmozhi, K. and Reddy, U.S.	An integrated machine vision and picture analysis device for insect detection enables improved identification of early-stage plant insects with a lower time and greater precision that will improve crop yield. Digital image processing methods were used in the current work to detect the insect forms in the sugar cane crop with photographs of crop insects for pre-processing, segmentation and extraction.	2017
15	Saha, A.K., Saha, J., Ray, R., Sircar, S., Dutta, S., Chattopadhyay, S.P. and Saha, H.N.	Drones with suitable cameras, sensors and modules can help to make agriculture simpler, more effective and more accurate. The solutions proposed relating to these drones will help to expand the potential of further improvement if combined with various machine learning and the Internet of Things concepts.	2018

S. No	Author name	Paper description	Publication year
16	Zhong, Y., Gao, J., Lei, Q. and Zhou, Y	The detection and coarse counting method on the basis of the detection and classification of objects by You Only Look Once (YOLO), using global functionalities is designed. Six insect species including fly, fruit fly, bee, mouth, moth, and chaff have been chosen to evaluate the efficacy of the system.	2018
17	Wang, Q.J., Zhang, S.Y., Dong, S.F., Zhang, G.C., Yang, J., Li, R. and Wang, H.Q	24 typical pest categories, which are mainly responsible for destroying field crops in China every year, are involved in Pest24. They use various cutting-edge methods of detection, such as RCNN Faster, SSD, YOLOv3 and Cascade R-CNN, to detect pests in the data set and to generate promising findings for in real time field pests monitoring.	2020
18	Ray, P.P	The devices and wireless communication technologies linked to IoT are evaluated extensively to meet the unique requirements of agricultural and agriculture applications. Research has been carried out on those IoT sensor systems which deliver intelligent and intelligent services to smart farming. Various case studies are provided to analyse the IoT-based solutions implemented according to their implementation parameters by different organisations and individuals and categories.	2017
19	Jiang, P., Chen, Y., Liu, B., He, D. and Liang, C.	This article is the first use of the Apple Leaf Disease Data Set (ALDD) data augmentation and annotation technology which consists of laboratory pictures and complex images in real life. Based on this proposal a new apple leaf detection model that uses deep-CNN is proposed by integrating the GoogLeNet Inception and Rainbow concatenation system.	2019
20	Behmann, J., Mahlein, A.K., Rumpf, T., Römer, C. and Plümer, L.	The findings relate both to the development of optical sensors that are non-invasive high resolution and methods for data processing that can manage signals from the resolution, size and complexity of these sensors. Several machinery learning methods, such as vector machines and neural classification (supervised learning) networks, have been used for precision agriculture (unsupervised learning).	2015
21	Rupanagudi, S.R., Ranjani, B.S., Nagaraj, P., Bhat, V.G. and Thippeswamy, G	The paper describes the methods for pesticide detection in one of the most popular tomato fruits in the world. "An insight into how in this project even the notion of the Internet of Things can be conceptualized".	2015
22	Nisha, G. and Megala, J.	A network of wireless sensors and temperature sensors distributed with solid moisture in the field is the unit. The Zigbee Protocol is used for the management of sensor information and for the control by means of an algorithm with sensor threshold values to a microcontroller irrigation system.	2014
23	Liu, H., Lee, S.H. and Chahl, J.S.	This paper is designed for exploring relative technologies and finding a method for the sensing and detection of crop invertebrates such as butterflies, sauté, snails and slugs. Two main areas for the identification and detection of invertebrates were identified: acoustic sensing and vision system (MVS).	2017

S. No	Author name	Paper description	Publication year
24	Kim, S., Lee, M. and Shin, C.	Cloud-based technology was built to manage information gathering, analysis and prediction in a shared framework. The proposed integrated system Farm as a Service (FaaS) assists high-level applications assistance through farms activity and observing the associated equipment, data and models management.	2018
25	Liu, H., Lee, S.H. and Chahl, J.S.	"They created a multi-spectral vision system that enables the development of denser plant and pest clouds with multi-spectral images of UV, blue, green, red, and near-infraround". An algorithm was designed to differentiate wide leafs from relatively large pests in nuclei at the noisy stage.	2017
26	Yan-e, D.	In this paper the methods for MIS intelligent agriculture design and architecture is discussed based on the implementation of the concept of management of agricultural knowledge and analysis of agricultural data characteristics.	2011

27. Conclusion

A literature review on IoT devices to identify and track insects in crop fields is discussed in this article. Each solution has been seen to have its possibilities and limitations. The species of pests must be increased and the identification must be improved. Detailed knowledge on the real-time and historical context is required to ensure effective control and allocation of capital. The Internet of Things allowed the rapid and efficient tracking of agricultural crops so as to increase the crop production and hence the farmer's revenues. In order to gather information on crop conditions and environmental changes, the wireless sensors network and sensors of various kinds were used.

This information is transmitted to the equipment network which initiates corrective action. For remote insect control and automatic insect identification, IoT and deep learning technologies are used. The Faster RCNN ResNet IoT Target Recognition Platform can be used to automatically identify the insect type with a four-layer IOT for remote trap insect monitoring.

Author details

Pooja Vajpayee[1*] and Kuldeep Kr. Yogi[2]

1 Department of Computer Science and Engineering, RKGIT, Ghaziabad, India

2 Department of Computer Science and Engineering, Banasthali Vidyapith, Jaipur, India

*Address all correspondence to: poojafcs@rkgit.edu.in

References

[1] The International Bank for Reconstruction and Development / The World Bank "ICT in agriculture - Connecting Smallholders to Knowledge, Networks, and Institutions" e-sourcebook, 2011

[2] Jim Chase: "The Evolution of the Internet of Things". White Paper, Texas Instruments, September, 2013.

[3] Patil, K.A. and Kale, N.R., 2016, December. "A model for smart agriculture using IoT". In *2016 International Conference on Global Trends in Signal Processing, Information Computing and Communication (ICGTSPICC)* (pp. 543-545). IEEE.

[4] Hongtao, Z., Bo, G. and Yuxia, H., 2011. "Design and realization of a real-time detection device for insect pests of field crops". In *Electrical Power Systems and Computers* (pp. 439-444). Springer, Berlin, Heidelberg.

[5] Liu, X.-y., Zhang, H.-t.: "Application of Fuzzy Recognition in the Detection of Pests in Fields". Journal of Agricultural Mechanization Research 31(12), 192-194 (2006)

[6] Zhao, H.-q., Shen, Z.-r., Yu, X.-w.: "On Computer-aided Insect Identification through Math-Morphology Features". Journal of China Agricultural University 7(3), 38-42 (2002)

[7] Drake, J.T., Walters, T.: "RITA, the Robot, Provides Survey and Identification Support to the USDA". Cphst News 2(9), 3-4 (2006)

[8] Shi, Y., Wang, Z., Wang, X. and Zhang, S., 2015, May. "Internet of things application to monitoring plant disease and insect pests". In *2015 International conference on Applied Science and Engineering Innovation*. Atlantis Press.

[9] Sreekantha, D.K. and Kavya, A.M., 2017, January. "Agricultural crop monitoring using IOT-a study". In *2017 11th International Conference on Intelligent Systems and Control (ISCO)* (pp. 134-139). IEEE.

[10] Ramalingam, B., Mohan, R.E., Pookkuttath, S., Gómez, B.F., Sairam Borusu, C.S.C., Wee Teng, T. and Tamilselvam, Y.K., 2020. "Remote Insects Trap Monitoring System Using Deep Learning Framework and IoT". *Sensors*, *20*(18), p.5280.

[11] Potamitis, I., Eliopoulos, P. and Rigakis, I., 2017. "Automated remote insect surveillance at a global scale and the internet of things". *Robotics*, 6(3), p.19.

[12] Materne, N. and Inoue, M., 2018, March. "Iot monitoring system for early detection of agricultural pests and diseases". In *2018 12th South East Asian Technical University Consortium (SEATUC)* (Vol. 1, pp. 1-5). IEEE.

[13] Deng, L., Wang, Y., Han, Z. and Yu, R., 2018. "Research on insect pest image detection and recognition based on bio-inspired methods". *Biosystems Engineering*, *169*, pp.139-148.

[14] Tirelli, P., Borghese, N.A., Pedersini, F., Galassi, G. and Oberti, R., 2011, May. "Automatic monitoring of pest insects traps by Zigbee-based wireless networking of image sensors". In *2011 IEEE International Instrumentation and Measurement Technology Conference* (pp. 1-5). IEEE.

[15] Nanni, L., Maguolo, G. and Pancino, F., 2020. Insect pest image detection and recognition based on bio-inspired methods. *Ecological Informatics*, p.101089.

[16] Cardim Ferreira Lima, M., Damascena de Almeida Leandro, M.E., Valero, C., Pereira Coronel, L.C. and Gonçalves Bazzo, C.O., 2020. Automatic Detection and Monitoring of Insect Pests—A Review. *Agriculture*, *10*(5), p.161.

[17] Miranda, J.L., Gerardo, B.D. and Tanguilig III, B.T., 2014. Pest detection and extraction using image processing techniques. *International Journal of Computer and Communication Engineering*, *3*(3), p.189.

[18] Xia, D., Chen, P., Wang, B., Zhang, J. and Xie, C., 2018. Insect detection and classification based on an improved convolutional neural network. *Sensors*, *18*(12), p.4169.

[19] Thenmozhi, K. and Reddy, U.S., 2017, November. Image processing techniques for insect shape detection in field crops. In *2017 International Conference on Inventive Computing and Informatics (ICICI)* (pp. 699-704). IEEE.

[20] Saha, A.K., Saha, J., Ray, R., Sircar, S., Dutta, S., Chattopadhyay, S.P. and Saha, H.N., 2018, January. IOT-based drone for improvement of crop quality in agricultural field. In *2018 IEEE 8th Annual Computing and Communication Workshop and Conference (CCWC)* (pp. 612-615). IEEE.

[21] Zhong, Y., Gao, J., Lei, Q. and Zhou, Y., 2018. A vision-based counting and recognition system for flying insects in intelligent agriculture. *Sensors*, *18*(5), p.1489.

[22] Wang, Q.J., Zhang, S.Y., Dong, S.F., Zhang, G.C., Yang, J., Li, R. and Wang, H.Q., 2020. Pest24: A large-scale very small object data set of agricultural pests for multi-target detection. *Computers and Electronics in Agriculture*, *175*, p.105585.

[23] Ray, P.P., 2017. Internet of things for smart agriculture: Technologies, practices and future direction. *Journal of Ambient Intelligence and Smart Environments*, *9*(4), pp.395-420.

[24] Jiang, P., Chen, Y., Liu, B., He, D. and Liang, C., 2019. Real-time detection of apple leaf diseases using deep learning approach based on improved convolutional neural networks. *IEEE Access*, *7*, pp.59069-59080.

[25] Behmann, J., Mahlein, A.K., Rumpf, T., Römer, C. and Plümer, L., 2015. A review of advanced machine learning methods for the detection of biotic stress in precision crop protection. *Precision Agriculture*, *16*(3), pp.239-260.

[26] Rupanagudi, S.R., Ranjani, B.S., Nagaraj, P., Bhat, V.G. and Thippeswamy, G., 2015, January. A novel cloud computing based smart farming system for early detection of borer insects in tomatoes. In *2015 International Conference on Communication, Information & Computing Technology (ICCICT)* (pp. 1-6). IEEE.

[27] Nisha, G. and Megala, J., 2014, December. Wireless sensor Network based automated irrigation and crop field monitoring system. In *2014 Sixth International Conference on Advanced Computing (ICoAC)* (pp. 189-194). IEEE.

[28] Liu, H., Lee, S.H. and Chahl, J.S., 2017. A review of recent sensing technologies to detect invertebrates on crops. *Precision Agriculture*, *18*(4), pp.635-666.

[29] Kim, S., Lee, M. and Shin, C., 2018. IoT-based strawberry disease prediction system for smart farming. *Sensors*, *18*(11), p.4051.

[30] Liu, H., Lee, S.H. and Chahl, J.S., 2017. A multispectral 3-D vision system for invertebrate detection on crops. *IEEE Sensors Journal*, *17*(22), pp.7502-7515.

[31] Yan-e, D., 2011, March. Design of intelligent agriculture management information system based on IoT. In *2011 Fourth International Conference on Intelligent Computation Technology and Automation* (Vol. 1, pp. 1045-1049). IEEE.